# Springer Series in Statistics

*Advisors:*
D. Brillinger, S. Fienberg, J. Gani,
J. Hartigan, K. Krickeberg

# Springer Series in Statistics

J. A. Hartigan

# Bayes Theory

Springer-Verlag
New York Berlin Heidelberg Tokyo

J. A. Hartigan

Department of Statistics
Yale University
Box 2179 Yale Station
New Haven, CT 06520
U.S.A.

AMS Classification: 62 A15

Library of Congress Cataloging in Publication Data

Hartigan, J. A.
    Bayes theory.
    (Springer series in statistics)
    Includes bibliographies and index.
    1. Mathematical statistics. I. Title. II. Series.
QA276.H392  1983     519.5     83-10591

With 4 figures.

Typeset by Thomson Press (India) Limited, New Delhi, India.

9 8 7 6 5 4 3 2 1

ISBN-13:978-1-4613-8244-7     e-ISBN-13:978-1-4613-8242-3
DOI: 10.1007/978-1-4613-8242-3

*To Jenny*

# Preface

This book is based on lectures given at Yale in 1971–1981 to students prepared with a course in measure-theoretic probability.

It contains one technical innovation—probability distributions in which the total probability is infinite. Such improper distributions arise embarrassingly frequently in Bayes theory, especially in establishing correspondences between Bayesian and Fisherian techniques. Infinite probabilities create interesting complications in defining conditional probability and limit concepts.

The main results are theoretical, probabilistic conclusions derived from probabilistic assumptions. A useful theory requires rules for constructing and interpreting probabilities. Probabilities are computed from similarities, using a formalization of the idea that the future will probably be like the past. Probabilities are objectively derived from similarities, but similarities are subjective judgments of individuals.

Of course the theorems remain true in any interpretation of probability that satisfies the formal axioms.

My colleague David Potlard helped a lot, especially with Chapter 13. Dan Barry read proof.

# Contents

# CHAPTER 1

# Theories of Probability

## 1.0. Introduction

A theory of probability will be taken to be an axiom system that probabilities must satisfy, together with rules for constructing and interpreting probabilities. A person using the theory will construct some probabilities according to the rules, compute other probabilities according to the axioms, and then interpret these probabilities according to the rules; if the interpretation is unreasonable perhaps the original construction will be adjusted.

To begin with, consider the simple finite axioms in which there are a number of elementary events just one of which must occur, events are unions of elementary events, and the probability of an event is the sum of the nonnegative probabilities of the elementary events contained in it.

There are three types of theory—logical, empirical and subjective. In logical theories, the probability of an event is the rational degree of belief in the event relative to some given evidence. In empirical theories, a probability is a factual statement about the world. In subjective theories, a probability is an individual degree of belief; these theories differ from logical theories in that different individuals are expected to have different probabilities for an event, even when their knowledge is the same.

## 1.1. Logical Theories: Laplace

The first logical theory is that of Laplace (1814), who defined the probability of an event to be the number of favorable cases divided by the total number of cases possible. Here cases are elementary events; it is necessary to identify equiprobable elementary events in order to apply Laplace's theory. In many

1

gambling problems, such as tossing a die or drawing from a shuffled deck of cards, we are willing to accept such equiprobability judgments because of the apparent physical indistinguishability of the elementary events—the particular face of the die to fall, or the particular card to be drawn. In other problems, such as the probability of it raining tomorrow, the equiprobable alternatives are not easily seen. Laplace, following Bernoulli (1713) used the principle of insufficient reason which specifies that probabilities of two events will be equal if we have no reason to believe them different. An early user of this principle was Thomas Bayes (1763), who apologetically postulated that a binomial parameter $p$ was uniformly distributed if nothing were known about it.

The principle of insufficient reason is now rejected because it sets rather too many probabilities equal. Having an unknown $p$ uniformly distributed is different from having an unknown $\sqrt{p}$ uniformly distributed, yet we are equally ignorant of both. Even in the gambling case, we might set all combination of throws of $n$ dice to have equal probability so that the next throw has probability 1/6 of giving an ace no matter what the results of previous throws. Yet the dice will always be a little biased and we want the next throw to have higher probability of giving an ace if aces appeared with frequency greater than 1/6 in previous throws.

Here, it is a consequence of the principle of insufficient reason that the long run frequency of aces will be 1/6, and this prediction may well be violated by the observed frequency. Of course any finite sequence will not offer a strict contradiction, but as a practical matter, if a thousand tosses yielded 1/3 aces, no gambler would be willing to continue paying off aces at 5 to 1. The principle of insufficient reason thus violates the skeptical principle that you can't be sure about the future.

## 1.2. Logical Theories: Keynes and Jeffreys

Keynes (1921) believed that probability was the rational belief in a proposition justified by knowledge of another proposition. It is not possible to give a numerical value to every such belief, but it is possible to compare some pairs of beliefs. He modified the principle of insufficient reason to a principle of indifference—two alternatives are equally probable if there is no relevant evidence relating to one alternative, unless there is corresponding evidence relating to the other. This still leaves a lot of room for judgment; for example, Keynes asserts that an urn containing $n$ black and white balls in unknown proportion will produce each sequence of white and black balls with equal probability, so that for large $n$ the proportion of white balls is very probably near 1/2. He discusses probabilities arising from analogy, but does not present methods for practical calculation of such probabilities. Keynes's theory does not succeed because it does not provide reasonable rules for computing probabilities, or even for making comparisons between probabilities.

Jeffreys (1939) has the same view of probability as Keynes, but is more constructive in presenting many types of prior distributions appropriate for different statistical problems. He presents an "invariant" prior distribution for a continuous parameter indexing a family of probability distributions, thus escaping one of the objections to the principle of insufficient reason. The invariant distribution is however inconsistent in another sense, in that it may generate conditional distributions that are not consistent with the global distribution. Jeffreys rejects it in certain standard cases.

Many of the standard prior probabilities used today are due to Jeffreys, and he has given some general rules for constructing probabilities. He concedes (1939, p. 37) that there may not be an agreed upon probability in some cases, but argues (p. 406) that two people following the same rules should arrive at the same probabilities. However, the many rules stated frequently give contradictory results.

The difficulty with Jeffreys's approach is that it is not possible to construct unique probabilities according to the stated rules; it is not possible to infer what Jeffreys means by probability by examining his constructive rules; it is not possible to interpret the results of a Jeffreys calculation.

## 1.3. Empirical Theories: Von Mises

Let $x_1, x_2, \ldots, x_n, \ldots$ denote an infinite sequence of points in a set. Let $f(A)$ be the limiting proportion of points lying in a set $A$, if that limit exists. Then $f$ satisfies the axioms of finite probability. In frequency theories, probabilities correspond to frequencies in some (perhaps hypothetical) sequence of experiments. For example "the probability of an ace is 1/6" means that if the same die were tossed repeatedly under similar conditions the limiting frequency would be 1/6.

Von Mises (1928/1964) declares that the objects under study are not single events but sequences of events. Empirically observed sequences are of course always finite. Some empirically observed sequences show approximate convergence of relative frequencies as the sample size increases, and approximate random order. Von Mises idealizes these properties in an infinite sequence or *collective* in which each elementary event has limiting frequency that does not change when it is computed on any subsequence in a certain family. The requirement of invariance is supposed to represent the impossibility (empirically observed) of constructing a winning betting system.

Non trivial collectives do not exist satisfying invariance over all subsequences but it is a consequence of the strong law of large numbers that

collectives exist that are invariant over any specified countable set of subsequences. Church (1940) suggests selecting subsequences using recursive
functions, functions of integer variables for which an algorithm exists that
will compute the value of the function for any values of the arguments in
finite time on a finite computing machine. There are countably many recursive functions so the collective exists, although of course, it cannot be
constructed. Further interesting mathematical developments are due to
Kolmogorov (1965) who defines a finite sequence to be random if an algorithm
required to compute it is sufficiently complex, in a certain sense; and to
Martin-Löf (1966) who establishes the existence of finite and infinite random
sequences that satisfy all statistical tests.

How is the von Mises theory to be applied? Presumably to those finite
sequences whose empirical properties of convergent relative frequency and
approximate randomness suggested the infinite sequence idealization. No
rules are given by von Mises for recognizing such sequences and indeed
he criticizes the "erroneous practice of drawing statistical conclusions from
short sequences of observations" (p. ix). However the Kolmogorov or
Martin-Lof procedures could certainly be used to recognize such sequences.

How does frequency probability help us learn? Take a long finite "random"
sequence of 0's and 1's. The frequency of 0's in the first half of the sequence
will be close to the frequency of 0's in the second half of the sequence, so that
if we know only the first half of the sequence we can predict approximately
the frequency of 0's in the second half, provided that we assume the whole
sequence is random. The prediction of future frequency is just a tautology
based on the assumption of randomness for the whole sequence.

It seems necessary to have a definition, or at least some rules, for deciding
when a finite sequence is random to apply the von Mises theory. Given such
a definition, it is possible to construct a logical probability distribution that
will include the von Mises limiting frequencies: define the probability of
the sequence $x_1, x_2, \ldots, x_n$ as $\lim_k N_k(x)/N_k$ where $N_k(x)$ is the number of
random sequences of length $k$ beginning with $x_1, x_2, \ldots, x_n$ and $N_k$ is the
number of random sequences of length $k$. In this way a probability is defined
on events which are unions of finite sequences. A definition of randomness
would not be acceptable unless $P[x_{n+1} = 1 \mid$ proportion of 1's in $x_1, \ldots,$
$x_n = p_n] - p_n \to 0$ as $n \to \infty$, that is, unless the conditional probability of
a 1 at the next trial converged to the limiting frequency of 1's.

True, definitions of randomness may vary, so that this is no unique solution—but the arbitrariness necessary to define finite randomness for applying
frequency theory is the same arbitrariness which occurs in defining prior
probabilities in the logical and subjective theories.

Asymptotically all theories agree; von Mises discusses only the asymptotic
case; to apply a frequency theory to finite sequences, it is necessary to make
the same kind of assumptions as Jeffreys makes on prior probabilities.

# 1.4. Empirical Theories: Kolmogorov

Kolmogorov (1933) formalized probability as measure: he *interpreted* probability as follows.

(1) There is assumed a complex of conditions $C$ which allows any number of repetitions.
(2) A set of elementary events can occur on establishment of conditions $C$.
(3) The event $A$ occurs if the elementary event which occurs lies in $A$.
(4) Under certain conditions, we may assume that the event $A$ is assigned a probability $P(A)$ such that
   (a) one can be practically certain that if the complex of conditions $C$ is repeated a large number of times $n$, then if $m$ be the number of occurrences of event $A$, the ratio $m/n$ will differ very slightly from $P(A)$.
   (b) if $P(A)$ is very small one can be practically certain that when conditions $C$ are realized only once, the event $A$ would not occur at all.

The axioms of finite probability will follow for $P(A)$, although the axiom of continuity will not.

As frequentists must, Kolmogorov is struggling to use Bernoulli's limit theorem for a sequence of independent identically distributed random variables without mentioning the word probability. Thus "the complex of conditions $C$ which allows any number of repetitions"—how different must the conditions be between repetitions? Thus "practically certain" instead of "with high probability." Logical and subjective probabilists argue that a larger theory of probability is needed to make precise the rules of application of a frequency theory.

# 1.5. Empirical Theories: Falsifiable Models

Statisticians in general have followed Kolmogorov's prescription. They freely invent probability models, families of probability distributions that describe the results of an experiment. The models may be falsified by repeating the experiment often and noting that the observed results do not concur with the model; the falsification, using significance tests, is itself subject to uncertainty, which is described in terms of the original probability model. A direct interpretation of probability as frequency appears to need an informal extra theory of probability (matching the circularity in Laplace's equally possible cases), but the "falsifiable model" interpretation appears to avoid the circularity. We propose a probability model, and then reject it, or modify it, if the observed results seem improbable. We are using Kolmogorov's rule (4)(b) that "formally" improbable results are "practically" certain not to happen. If they do happen we doubt the formal probability.

The weaknesses in the model approach:

(1) The repetitions of the experiment are assumed to give independent, identically distributed results. Otherwise laws of large numbers will not apply. But you can't test that independence without taking some other series of experiments, requiring other assumptions of independence, and requiring other tests. In practice the assumption of independence is usually untested (often producing very poor estimates of empirical frequencies; for example, in predicting how often a complex piece of equipment will break, it is dangerous to assume the various components will break independently). The assumption of independence in the model theory is the analogue of the principle of insufficient reason in logical theories. We assume it unless there is evidence to the contrary, and we rarely collect evidence.

(2) Some parts of the model, such as countable additivity or continuity of a probability density, are not falsifiable by any finite number of observations.

(3) Arbitrary decisions about significance tests must be made; you must decide on an ordering of the possible observations on their degree of denial of the model—perhaps this ordering requires subjective judgment depending on past knowledge.

## 1.6. Subjective Theories: De Finetti

De Finetti (1930/1937) declares that the degree of probability attributed by an individual to a given event is revealed by the conditions under which he would be disposed to bet on that event. If an individual must bet on all events $A$ which are unions of elementary events, he must bet according to some probability $P(A)$ defined by assigning non-negative probabilities to the elementary events, or else a dutch book can be made against him—a combination of bets is possible in which he will lose no matter which elementary event occurs. (This is only a little bit like von Mises's principle of the impossibility of a gambling system.) De Finetti calls such a system of bets coherent.

In the subjectivist view, probabilities are associated with an individual. Savage calls them "personal" probabilities; a person should be coherent, but any particular event may be assigned any probability without questioning from others. You cannot say that "my probability that it will rain this afternoon is .97" is wrong—it reports my willingness to bet at a certain rate. Bayes (1763) defines probability as "the ratio between the value at which an expectation depending on the happening of the event ought to be computed, and the value of the thing expected upon its happening." His probability describes how a person *ought* to bet, not how he *does* bet. It should be noted that the subjectivist theories insist that a person be coherent in his betting, so that they are not content to let a person bet how he pleases; psychological probability comes from the study of actual betting behavior, and indeed people are consistently incoherent (Wallsten (1974)).

There are numerous objections to the betting approach some technical (is it feasible?), others philosophical (is it useful?).

(i) *People don't wish to offer precise odds*—Smith (1961) and others have suggested ranges of probabilities for each event; this is not a very serious objection.

(ii) *A bet is a price, subject to market forces*—depending on the other actors; Borel (1924) considers the case of a poker player, who by betting high, increases his probability of winning the pot. Can you say to him "your probability of winning the pot is the amount you are willing to bet to win the pot divided by the amount in the pot."

Suppose you are in a room full of knowledgeable meteorologists, and you declare the probability it will rain tomorrow is .95. They all rush at you waving money. Don't you modify the probability? We may not be willing to bet at all if we feel others know more. Why should the presence of others be allowed to affect our probability?

(iii) *The utility of money is not linear*—You may bet $1 to win $500 when the chance of winning is only 1/1000; the gain of $500 seems more than 500 times the loss of $1. Ramsey (1926) and Savage (1954) advance theories of rational decision making, choosing among a range of available actions, that produce both utilities and probabilities for which the optimal decision is always that decision which maximizes expected utility.

The philosophical objection is that *I* don't particularly care how *you* (opinionated and uninformed as you are) wish to bet. To which the subjectivists will answer that subjective judgments are necessary in forming conclusions from observations; let us be explicit about them (Good (1976, p. 143)). To which the empiricists will reply, let us separate the "good" empirically verifiable probabilities, the likelihoods, from the "bad" subjective probabilities which vary from person to person. (Cox and Hinkley (1974, p. 389) "For the initial stages ... the approach is ... inapplicable because it treats information derived from data as on exactly equal footing with probabilities derived from vague and unspecified sources.")

## 1.7. Subjective Theories: Good

Good (1950) takes a degree of belief in a proposition $E$ given a proposition $H$ and a state of mind of a person $M$, to be a primitive notion allowing no precise definition. Comparisons are made between degrees of belief; a set of comparisons is called a *body of beliefs*. A *reasonable* body of beliefs contains no contradictory comparisons.

The usual axioms of probability are assumed to hold for a numerical probability which has the same orderings as a body of beliefs. Good recommends a number of rules for computing probabilities, including for example the device of imaginary results: consider a number of probability assignments to a certain event; in combination with other fixed probability judgments, each will lead through the axioms to further probability judgments; base your original choice for probabilities on the palatability of the overall probabilities which ensue. If an event of very small probability occurs, he suggests that the body of beliefs be modified.

Probability judgments can be sharpened by laying bets at suitable odds, but there is no attempt to define probability in terms of bets. Good (1976, p. 132) states that "since the degrees of belief, concerning events over which he has no control, of a person with ideally good judgment, should surely not depend on whether he uses his beliefs in any specific manner, it seems desirable to have justifications that do not mention preferences or utilities. But utilities necessarily come in whenever the beliefs are to be used in a practical problem involving action."

Good takes an attitude, similar to the empirical model theorists, that a probability system proposed is subject to change if errors are discovered through significance testing. In standard probability theory, changes in probability due to data take place according to the rules of conditional probability; in the model theory, some data may invalidate the whole probability system and so force changes not according to the laws of probability. There is no contradiction in following this practice because we separate the formal theory from the rules for its application.

## 1.8. All the Probabilities

An overview of the theories of probability may be taken from the stance of a subjective probabilist, since subjective probability includes all other theories. Let us begin with the assumption that an individual attaches to events numerical probabilities which satisfy the axioms of probability theory.

If no rules for constructing and interpreting probabilities are given, the probabilities are inapplicable—for all we know the person might be using length or mass or dollars or some other measure instead of probability. Thus the theories of Laplace and Keynes are not practicable for lack of rules to construct probability. Jeffreys provides rules for many situations (although the rules are inconsistent and somewhat arbitrary). Good takes a belief to be a primitive notion; although he gives numerous rules for refining and correcting sets of probabilities, I believe that different persons might give different probabilities under Good's system, on the same knowledge, simply because they make different formalizations of the primitive notion of degree of belief. Such disagreements are accepted in a subjective theory, but it seems undesirable that they are caused by confusion about meanings of probability. For example if you ask for the probability that it will rain tomorrow afternoon, one person might compute the relative frequency of rain on afternoons in the last month, another might compute the relative amount of today's rain that fell this afternoon; the axioms are satisfied. Are the differences in computation due to differences in beliefs about the world, or due to different interpretations of the word probability?

The obvious interpretation of a probability is as a betting ratio, the amount you bet over the amount you get. There are certainly some complications in this interpretation—if a probability is a price, it will be affected by the market in which the bet is made. But these difficulties are overcome by Savage's treatment of probability and utility in which an individual is asked to choose coherently between actions, and then must do so to maximize expected utility as measured by an implied personal probability and utility. The betting interpretation arises naturally out of the foundations of probability theory as a guide to gamblers, and is not particularly attached to any theory of probability. A logical probabilist, like Bayes, will say that a probability is what you *ought* to bet. A frequentist will say that a bet is justified only if it would be profitable in the long run—Fisher's evaluation of estimation procedures rests on which would be more profitable in the long run. A subjectivist will say that the probability is the amount *you* are willing to bet, although he will require coherence among your bets. It is therefore possible to adopt the betting interpretation without being committed to a particular theory of probability.

As Good has said, the frequency theory is neither necessary nor sufficient. Not sufficient because it is applicable to a single type of data. Not necessary because it is neatly contained in logical or subjectivist theories, either through Bernoulli's celebrated law of large numbers which originally generated the frequency theory, or through de Finetti's celebrated convergence of conditional probabilities on exchangeable sequences, which makes it clear what probability judgments are necessary to justify a frequency theory. (A sequence $x_1, x_2, \ldots, x_n, \ldots$ is exchangeable if its distribution is invariant under finite permutations of the indices, and then if the $x_i$ have finite second moment, the expected value of $x_{n+1}$ given $x_1, \ldots, x_n$ and $(1/n)\sum x_i$ converge to the same limiting random variable.) Thus the frequency theory gives an approximate value to conditional expectation for data of this type: the sequence of repeated experiments must be judged exchangeable.

The frequency theory does not assist with the practical problem of prediction from short sequences. Nor does it apply to other types of data. For example we might judge that the series is stationary rather than exchangeable: the assumption is weaker but limit results still apply under certain conditions. The frequency theory would be practicable if data consisted of long sequences of exchangeable random variables (the judgment of exchangeability being made informally, outside the theory); but too many important problems are not of this type.

The model theory of probability uses probability models that are "falsified" if they give very small probability to certain events. The only interpretation of probability required is that events of small probability are assumed "practically certain" not to occur. The advance over the frequency theory is that it is not necessary to explain what repeatable experiments are. The loss is that many probabilities must be assumed in order to compute the probabilities of the falsifying events, and so it is not clear which probabilities are false if one of the events occur. The interpretation of small probabilities as practically zero is not adequate to give meaning to probability. Consider for example the model that a sample of $n$ observations is independently sampled from the normal: one of the observations is 20 standard deviations from the rest; we might conclude that the real distribution is not normal or that the sampled observations are not independent (for example the first $(n - 1)$ observations may be very highly correlated). Thus we cannot empirically test the normality unless we are sure of the independence; and assuming the independence is analogous to assuming exchangeability in de Finetti's theories.

Finally the subjective theory of probability is objectionable because probabilities are mere personal opinions: one can give a little advice; the probabilities should cohere, the set of probabilities should not combine to give unacceptable probabilities; but in the main the theory *describes* how ideally rational people act rather than *recommends* how they should act.

## 1.9. Infinite Axioms

Two questions arise when probabilities are defined on infinite numbers of events. These questions cannot be settled by reference to empirical facts, or by considering interpretations of probability, since in practice we do not deal with infinite numbers of events. Nevertheless it makes a considerable difference in the mathematics which choices are made.

In Kolmogorov's axioms, the axiom of countable additivity is assumed. This makes it possible to determine many useful limiting probabilities that would be unavailable if only finite additivity is assumed, but at the cost of limiting the application of probability to a subset of the family of all subsets of the line. Philosophers are reluctant to accept the axiom, but mathematicians are keen to accept it; de Finetti and others have developed a theory of finitely additive probability which differs in exotic ways from the regular theories—he will say "consider the uniform distribution on the line, carried by the rationals"; distribution functions do not determine probability distributions on the line. Here, the axiom of countable additivity is accepted as a mathematical convenience.

The second infinite axiom usually accepted is that the total probability should be one. This is inconvenient in Bayes theory because we frequently need uniform distributions on the line; countable additivity requires that total probability be infinite.

Allowing total probability to be infinite does not prevent interpretation in any of the standard theories. Suppose probability is defined on a suitable class of functions $X$. Probability judgments may all be expressed in the form $PX \geq 0$ for various $X$. In the frequency theory, given a sequence $x_1, x_2, \ldots,$ $x_n, \ldots PX \geq 0$ means that $\sum_{i=1}^{n} x_i \geq 0$ for all large $n$. In the betting theory, $PX \geq 0$ means that you are willing (subjective) or ought (logical) to accept the bet $X$.

## 1.10. Probability and Similarity

I think there is probability about 0.05 that there will be a large scale nuclear war between the U.S. and the U.S.S.R. before 2000. By that I certainly don't mean that such nuclear exchanges will occur in about one in twenty of some hypothetical infinite sequence of universes. Nor do I mean that I am willing to bet on nuclear war at nineteen to one odds—I am willing to accept any wager that I don't have to pay off until after the bomb. (I trust the U.S.S.R. targeting committee to have put aside a little something for New Haven, and even if they haven't, bits of New York will soon arrive by air.)

What then does the probability 0.05 mean? Put into an urn 100 balls differing only in that 5 are black and 95 are white. Shake well and draw a ball without looking. I mean that the probability of nuclear war is about the same as the probability of getting a black ball (or more precisely, say, war is more probable than drawing a black ball when the urn has 1 black and 99 white balls, and less probable than drawing a black ball when the urn has 10 black and 90 white balls.) You might repeat this experiment many times and expect 5% black balls, and you might be willing to bet at 19 to 1 that a black ball will appear, although of course the decision to bet will depend on other things such as your fortune and ethics. To me, the probability .05 is meaningful for the 5 out of 100 balls indistinguishable except for color, without reference to repeating the experiment or willingness to bet.

Why should you believe the assessment of .05? I need to offer you the data on which the probability calculation is based. The superpowers could become engaged in a nuclear war in the following ways.

1. *Surprise attack.* Such an attack would seem irrational and suicidal; but war between nations has often seemed irrational and suicidal. For example the Japanese attack on the United States in 1941 had a good chance of resulting in the destruction of the Japanese Empire, as the Japanese planners knew, but they preferred chancy attack to what they saw as sure slow economic strangulation. Might not the U.S. or the U.S.S.R. attack for similar reasons? If such an attack occurs once in 1000 years, the chance of it occurring in the next twenty is .02 (I will concede that the figure might be off by a factor of 10.)

2. *Accidental initiation.* Many commanders have the physical power to

launch an attack, because the strategic systems emphasize fast response times to a surprise attack. Let us say 100 commanders, and each commander so stable that he has only one chance in 100,000 of ordering a launch, in a given year, and that the isolated launch escalates to a full scale exchange with probability .2. In twenty years, a nuclear war occurs with probability one .004.

3. *Computer malfunction.* Malfunctions causing false alerts have been reported at least twice in the U.S. press in the last twenty years. Let us assume that a really serious malfunction causing a launch is 100 times as rare. In the next twenty years we expect only .02 malfunctions.

4. *Third party initiation.* Several embattled nations have nuclear capability—Israel (.99 probability), South Africa (.40), India (.60), Pakistan (.40), Libya (.20). Of these Israel is the most threatened and the most dangerous. Who would be surprised by a preemptive nuclear attack by Israel on Libyan nuclear missile sites? Let us say the probability is .01, and the chance of escalation to the superpowers is .2. The overall probability is .002.

Summing the probabilities we get .046, say .05, which I admit may be off by a factor of 10. There is plenty of room for disagreement about the probabilities used in the calculations; and indeed I have committed apparently a circular argument characteristic of probability calculations; I am supposed to be showing how a probability is to be calculated, but I am basing the calculation on other probabilities. How are *they* to be justified?

The component probabilities are empirical, based on occurrences of similar events to the one being assessed. An attack by the U.S. on the U.S.S.R. is analogous to the attack by Japan on the U.S. Dangerously deceptive computer malfunctions have already occurred. Of course the analogies are not very close, because the circumstances of the event considered are not very similar to the circumstances of the analogous events.

The event of interest has been expressed as the disjoint union of the intersections of "basic" events (I would like to call them atomic events but the example inhibits me!). Denote a particular intersection as $B_1 B_2 \ldots B_n$. The probability $P(B_1 B_2 \ldots B_n) = P(B_1)P(B_2 | B_1) \ldots P(B_n | B_1 B_2 \ldots B_{n-1})$ is computed as a product of conditional probabilities. The conditional probability $P(B_i | B_1 B_2 \ldots B_{i-1})$ is computed from the occurrence of events similar to $B_i$ under conditions similar to $B_1 B_2 \ldots B_{i-1}$. We will feel more or less secure in the probability assessment according to the degree of similarity of the past events and conditions to $B_i$ and $B_1 B_2 \ldots B_{i-1}$. The probability calculations have an objective, empirical part, in the explicit record of past events, but also a subjective judgmental part, in the selection of "similar" past events. Separate judgments are necessary in expressing the event of interest in terms of basic events—we will attempt to use basic events for which a reasonable empirical record exists.

The future is likely to be like the past. Probability must therefore be a function of the similarities between future and past events. The similarities will be subjective, but given the similarities a formal objective method should be possible for computing probabilities.

# 1.11. References

Bayes, T. (1763), An essay towards solving a problem in the doctrine of chances, *Phil. Trans. Roy. Soc.* **53**, 370–418, **54**, 296–325, reprinted in *Biometrika* **45** (1958), 293–315.

Bernoulli, James (1713), Ars Conjectandi.

Borel, E. (1924), Apropos of a treatise on probability, *Revue philosophique*, reprinted in H. E. Kyburg and H. E. Smokler (eds.), *Studies in Subjective Probability*. London: John Wiley, 1964, pp. 47–60.

Church, A. (1940), On the concept of a random sequence, *Bull. Am. Math. Soc.* **46**, 130–135.

Cox, D. R. and Hinkley, D. V. (1974), *Theoretical Statistics*. London: Chapman and Hall.

De Finetti, B. (1937), Foresight: Its logical laws, in subjective sources, reprinted in H. E. Kyburg and H. E. Smokler (eds.), *Studies in Subjective Probability*. London: John Wiley, 1964, pp. 93–158.

Good, I. J. (1950), *Probability and the Weighing of Evidence*. London: Griffin.

Good, I. J. (1976), The Bayesian influence, or how to sweep subjectivism under the carpet, in Harper and Hooker (eds.), *Foundations of Probability Theory, Statistical Inference, and Statistical Theory of Science*. Dordrecht: Reidel.

Jeffreys, H. (1939), *Theory of Probability*. London: Oxford University Press.

Keynes, J. M. (1921), *A Treatise on Probability*. London: MacMillan.

Kolmogorov, A. N. (1950). *Foundations of the Theory of Probability*. New York: Chelsea. (The German original appeared in 1933.)

Kolmogorov, A. N. (1965), Three approaches to the quantitative definition of information, *Problemy Peredaci Informacii* **1**, 4–7.

Laplace, P. S. (1814), *Essai philosophique sur les probabilités*, English translation. New York: Dover.

Martin-Löf, M. (1966), The definition of random sequences, *Information and Control* **9**, 602–619.

Ramsey, F. (1926), Truth and probability, reprinted in H. E. Kyburg and H. E. Smokler (eds.), *Studies in Subjective Probability*. New York: John Wiley, 1964, pp. 61–92.

Savage, L. J. (1954), *The Foundations of Statistics*. New York: John Wiley.

Smith, C. A. B. (1961). Consistency in statistical inference and decision, *J. Roy. Statist. Soc. B* **23**, 1–25.

von Mises, R. and Geiringer, H. (1964), *The Mathematical Theory of Probability and Statistics*. New York: Academic Press.

Wallsten, Thomas S. (1974), The psychological concept of subjective probability: a measurement theoretic view: in C. S. Staël von Holstein (ed.), *The Concept of Probability in Psychological Experiments*. Boston: Reidel, p. 49–72.

# CHAPTER 2

# Axioms

## 2.0. Notation

The objects of probability will be bets $X$, $Y$, ... that have real-valued payoffs $X(s)$, $Y(s)$, ... according to the true state of nature $s$, where $s$ may be any of the states in a set $S$.

Following de Finetti, events will be identified with bets taking only the values 0 and 1. In particular, the notation $\{s$ satisfies certain conditions$\}$ will denote the event equal to 1 when $s$ satisfies the conditions, and equal to 0 otherwise. For example $\{X \leq 5\}$ denotes the event equal to 1 when $s$ is such that $X(s) \leq 5$, and equal to 0 otherwise.

In general algebraic symbols $+$, $-$, $\leq$, $\vee$, $\wedge$ will be used rather than set theoretic symbols $\cup$, $\cap$, $\subset$.

## 2.1. Probability Axioms

Let $S$ denote a set of outcomes, let $X$, $Y$, ... denote bets on $S$, real valued functions such that $X(s)$, $Y(s)$, ... are the payoffs on the bets when $s$ occurs. A *probability space* $\mathscr{X}$ is a set of bets such that

(1) $X, Y \in \mathscr{X} \Rightarrow aX + bY \in \mathscr{X}$   for $a$, $b$ real
(2) $X \in \mathscr{X} \Rightarrow |X| \in \mathscr{X}$
(3) $X \in \mathscr{X} \Rightarrow X \wedge 1 \in \mathscr{X}$
(4) $|X_n| \leq X_0 \in \mathscr{X}$,    $X_n \to X \Rightarrow X \in \mathscr{X}$.

A *probability* $P$ on $\mathscr{X}$ is a real valued function on $\mathscr{X}$ that is

LINEAR: $P(aX + bY) = aPX + bPY$   for $X$, $Y \in \mathscr{X}$ and $a$, $b$ real
NON-NEGATIVE: $P|X| \geq 0$   for $X \in \mathscr{X}$
CONTINUOUS: $|X_n| \leq X_0 \in \mathscr{X}$,    $X_n \to X \Rightarrow PX_n \to PX$.

14

A *unitary probability* $P$ is defined on a probability space $\mathscr{X}$ such that $1 \in \mathscr{X}$, and satisfies $P1 = 1$. A *finitely additive* probability $P$ is defined on a linear space such that $X \in \mathscr{X} \Rightarrow |X| \in \mathscr{X}$, $1 \in \mathscr{X}$, and $P$ is linear and non-negative, but not necessarily continuous.

A probability space $\mathscr{X}$ is *complete* with respect to $P$ if

(i)   $X_n \in \mathscr{X}, \qquad P|X_n - X_m| \to 0, \qquad X_n \to X \Rightarrow X \in \mathscr{X}$
(ii)  $Y \in \mathscr{X}, \qquad 0 \leqq X \leqq Y, \qquad PY = 0 \ \Rightarrow X \in \mathscr{X}.$

The standard definition of probability, set down in Kolmogorov (1933), requires that it be unitary. According to Keynes (1921, p. 155), it was Leibniz who first suggested representing certainty by 1. However, in Bayes theory it is convenient to have distributions which have $P1 = \infty$, such as the uniform distributions over the line and integers. Jeffreys allows $P1 = \infty$, because his methods of generating prior distributions frequently produce such $P$, but in theoretical work with probabilities, he usually assumes $P1 = 1$. Renyi (1970) handles infinite probabilities using families of conditional probabilities. But there is no formal theory to handle probabilities with $P1 = \infty$, which are therefore called *improper*. The measure theory for this case is well developed; see for example, Dunford and Schwartz (1964).

The betting theory interpretation of probability is straightforward; $PX_1/PX_2$ is the relative value of bet $X_1$ to bet $X_2$; you accept only bets $X$ such that $PX \geqq 0$. It is true that you may effectively give value $\infty$ to the constant bet 1; those bets which you wish to compare are infinitely less valuable than 1.

It is also possible to make a frequency interpretation of non-unitary probability. Consider for example the uniform distribution over the integers. This would be the limiting frequency probability of a sequence of integers, such as

$$1 \quad 12 \quad 123 \quad 1234 \quad 12345 \quad 123456\ldots$$

in which each pair of integers occurred with the same limiting relative frequency. If it is insisted that continuity hold, then total probability is infinite. If it is insisted that total probability is 1, then continuity breaks down and the limiting frequency probability is finitely additive.

It is not possible to justify either the continuity axiom or probabilities with $P1 = \infty$ by reference to actual experience, which is necessarily finite. Indeed de Finetti rejects the continuity axiom on this and other grounds. But the continuity axiom equally cannot be denied by reference to experience, and it is mathematically convenient in permitting *unique* extension of $P$ defined on some small set of functions to $P$ defined on a larger set of interesting limit functions: we begin by assuming that intervals have probability proportional to their length, and end by stating that the rationals have probability zero. [In contrast, de Finetti (1972) can say: consider the uniform distribution on the real line carried by the rationals, or carried by the irrationals.] We need to invent methods to handle invented concepts such as the set of rationals; the main justification must be mathematical convenience; and the same reasoning applies to non-unitary probabilities – they must be mathematically convenient or they would not be so improperly ubiquitous (see them used by de Finetti, 1970, p. 237).

## 2.2. Prespaces and Rings

A *prespace* $\mathscr{A}$ is a linear space such that $X \in \mathscr{A} \Rightarrow |X| \in \mathscr{A}, X \wedge 1 \in \mathscr{A}$. A *limit space* $L$ is such that $X_n \in L, X_0 \in L, |X_n| \leq X_0, X_n \to X$ implies $X \in L$. A probability space is both a prespace and a limit space.

**Lemma.** *The smallest probability space including a prespace $\mathscr{A}$ is the smallest limit space including $\mathscr{A}$.*

PROOF. Let $L$ be the intersection of limit spaces, containing $\mathscr{A}$. For each $X$, let $L(X)$ be the set of functions $Y$ such that $\mathscr{A}(X, Y): |X|, |Y|, X \wedge 1, Y \wedge 1, aX + bY$ lie in $L$; $L(X)$ is a limit space. If $X \in \mathscr{A}$, then $\mathscr{A}(X, Y) \subset \mathscr{A} \subset L$ for $Y$ in $\mathscr{A}$. Thus $L(X) \supset \mathscr{A} \Rightarrow L(X) \supset L$. If $X \in L$, then $X \in L(Y) \supset L$ for $Y$ in $\mathscr{A}$. Thus $L(X) \supset \mathscr{A} \Rightarrow L(X) \supset L$. If $X \in L$, $Y \in L$ then $\mathscr{A}(X, Y) \subset L$, so $L$ is a prespace and therefore a probability space.                                                    □

A probability $P$ on a prespace $\mathscr{A}$ is linear, non-negative and continuous: $X_n \to 0, |X_n| \leq X \in \mathscr{A} \Rightarrow PX_n \to 0$.

**Theorem.** *A probability $P$ on a prespace $\mathscr{A}$ may be uniquely extended to a probability $P$ on a completed probability space including $\mathscr{A}$.*

PROOF. Let $\mathscr{X}$ consist of functions $X$ for which $|X - a_n| \leq \sum_{i=1}^{\infty} a_n^i, a_n^i \geq 0$, $\sum_{i=1}^{\infty} Pa_n^i \to 0$ where $a_n, a_n^i \in \mathscr{A}$. Say that the sequence $a_n$ approximates $X$ and define $PX = \lim_n Pa_n$. It follows from continuity that the definition is unique, which implies that $P$ is unchanged for $X$ in $\mathscr{A}$. If $a_n, b_n$ approximate $X, Y$ then $aa_n + bb_n$ approximates $aX + bY$ with $P(aX + bY) = aPX + bPY$. $|a_n|$ approximates $|X|$ with $P|X| \geq 0$, and $a_n \wedge 1$ approximates $X \wedge 1$.

Now suppose $X_n \in \mathscr{X}, X \in \mathscr{X}, |X_n| \leq X$ and $X_n \to Y$. We will show that $Y \in \mathscr{X}$ and $PY = \lim_n PX_n$. First assume $X_n \uparrow Y$. Then $P|X_{n+1} - X_n| < 2^{-n}$ on a suitably chosen subsequence. Also $|X_{n+1} - X_n| \leq \sum_{i=1}^{\infty} a_n^i$ where $a_n^i \in \mathscr{A}$, $a_n^i \geq 0$ and $\sum Pa_n^i < 2^{-n+1}$, since $|X_{n+1} - X_n| \in \mathscr{X}$. Thus $|Y - X_n| \leq \sum_{i=n}^{\infty} |X_{i+1} - X_i| \leq \sum_{i=n}^{\infty} \sum_{j=1}^{\infty} a_i^j$ where $\sum \sum Pa_i^j \leq 2^{-n+2}$. Approximate $X_n$ by $a_n$ where $P|X_n - a_n| < 2^{-n+2}$. Then $Y$ is approximated by $a_n$, and $PY = \lim Pa_n = \lim PX_n$. The general result follows using

$$\sup_{N \leq n \leq M} X_n \uparrow \sup_{N \leq n} X_n \quad \text{and} \quad \sup_{N \leq n} X_n \downarrow Y.$$

If $X_n \in \mathcal{X}$, $X_n \to X$ and $P|X_n - X_m| \to 0$, a similar argument, first considering monotone convergence, shows that $X \in \mathcal{X}$. If $a_n$ approximates $Y$, it approximates $X, 0 \leq X \leq Y$, so $\mathcal{X}$ is complete with respect to $P$.

Suppose $P'$ is a probability on $\mathcal{X}$ which agrees with $P$ on $\mathcal{A}$. Then $P'|X - a_n| \to 0$ if $a_n$ approximates $X$, so $P'X = \lim P'a_n = \lim Pa_n = PX$. Thus $P$ is uniquely defined on $\mathcal{X}$. $\qquad\qquad\square$

A subset of $S$ is a function on $S$ taking the values 0 and 1.

A family $\mathcal{F}$ of subsets of $S$ is a *ring* if $A, B \in \mathcal{F} \Rightarrow A \cup B, A - AB \in \mathcal{F}$.

A function $P$ on $\mathcal{F}$ is a *probability* if

(i)    $P(A + B) = PA + PB$    if $A, B \in \mathcal{F}$, $AB = 0$

(ii)    $PA \geq 0$    for $A$ in $\mathcal{F}$

(iii)   $A_n \to 0$,   $A_n \leq A \in \mathcal{F} \Rightarrow PA_n \to 0$.

(iii)'   $A_n \downarrow 0 \Rightarrow PA_n \downarrow 0$.

[Note that (iii) and (iii)' are equivalent. Obviously (iii) $\Rightarrow$ (iii)'. Suppose that (iii)' holds, and $A_n \to 0$, $A_n < A$. Define $B_n = A_n \cup A_{n+1} \ldots \cup A_m$ where $2^{-n} + PB_n > \sup\limits_m P(A_{n+1} \cup \ldots \cup A_m) \leq PA$. Then

$$P[B_m - B_m B_n] = P[B_m \cup B_n - B_n] < 2^{-n} \quad \text{for } m > n$$

$$P[B_m B_n - B_m B_n B_{n+1}] \leq P[B_m - B_m B_{n+1}] \leq 2^{-(n+1)} \quad \text{for } m > n+1$$

$$P[B_m - \prod_{n \leq i \leq m} B_i] \leq 2^{-n+1}$$

Since $\prod_{n \leq i \leq m} B_i \downarrow 0$ as $m \to \infty$, $P(\prod B_i) \downarrow 0$, so $\overline{\lim} \, PB_m \leq 2^{-n+1}$ for every $n$. Since $A_m \leq B_m$, $\lim PA_m < 2^{-n+1}$ for every $n$, $PA_m \to 0$.]

If $P$ is a probability on $\mathcal{F}$ it may be uniquely extended to the prespace $\mathcal{A}$ consisting of elements $\sum_{i=1}^n \alpha_i A_i$, where $\alpha_i$ is real, by $P(\sum_{i=1}^n \alpha_i A_i) = \sum_{i=1}^n \alpha_i PA_i$. It is easily checked that $P$ is well defined, linear and non-negative on $\mathcal{A}$. The continuity condition is a little more difficult; suppose $a_n \to 0$, $|a_n| \leq a$. Then $|a_n| \leq \lambda A$ for some positive $\lambda$, $A \in \mathcal{F}$.

$$|a_n| \leq \varepsilon A + \{|a_n| \geq \varepsilon\} \lambda A.$$

Since $\{|a_n| \geq \varepsilon\} A \to 0$ and $\{|a_n| \geq \varepsilon\} A \leq A$, $P\{|a_n| \geq \varepsilon\} \lambda A \to 0$. Thus $\overline{\lim} \, P|a_n| \leq \varepsilon PA$ for every $\varepsilon > 0$ and $P|a_n| \to 0$ as $n \to \infty$.

If $P$ is a probability on $\mathcal{F}$ it may, by Theorem 2.2, be extended uniquely to the smallest complete probability space $\mathcal{X}$ including $\mathcal{F}$. It is customary to call $P$ on $\mathcal{X}$ an expectation or an integral, but we follow de Finetti in identifying sets and functions, probabilities and expectations.

If $P$ is a probability on $\mathcal{X}$, $P$ defined on the ring $\mathcal{F}$ of $0-1$ functions in $\mathcal{X}$ extends uniquely to a complete probability space $\mathcal{X}(\mathcal{F})$ that includes $\mathcal{X}$. See Loomis (1953). Thus specifying $P$ on $\mathcal{F}$ determines it on $\mathcal{X}$; the function $X$ is approximated by the step functions

$$\sum_{1 < |k| < K^2} \frac{k}{K} \left\{ \frac{k}{K} \leq X < \frac{k+1}{K} \right\}$$

and

$$PX = \lim_{K \to \infty} \sum \frac{k}{K} P\left\{ \frac{k}{K} \leqq X < \frac{k+1}{K} \right\}.$$

EXAMPLE. Let $\mathscr{F}$ be the set of finite unions of half-open intervals $A = U(a_i, b_i]$. Define $PA = \sum |b_i - a_i|$ if the intervals $(a_i, b_i]$ are disjoint. To check that $P$ is a probability, it is difficult only to prove (iii)'. Assume $A_n \downarrow 0$.

Let $A_n = U_{i=1}^n (a_{in}, b_{in}]$, and if $A_n < A$, let $A$ be the interval $[a, b]$. The function $A - A_n$ is a union of half open intervals which converges to $A$. Define $E = \underset{n}{U} \underset{i}{U} (a_{in} - (\varepsilon/2^n), a_{in} + (\varepsilon/2^n))$. Then $(A - A_n) \cup E$ is an open set, and $U(A - A_n) \cup E$ includes $[a, b]$. From compactness, a finite number of $(A - A_n) \cup E$ cover $[a, b]$, and since $A - A_n \uparrow$, for some $n$, $(A - A_n) \cup E \supset A$. Since $E$ has total length less than $\varepsilon$, $A_n$ must have total length less than $\varepsilon$. Thus $PA_n \to 0$.

From length of intervals on $\mathscr{F}$, we define a probability on a prespace of step functions on intervals; from the prespace we define probabilities on a probability space $\mathscr{X}$ which includes, for example, all continuous functions zero outside a finite interval. This is lebesgue measure.

## 2.3. Random Variables

Let $P$ be a probability on $\mathscr{Y}$ a probability space on $T$, and let $\mathscr{X}$ be a probability space on $S$. A *random variable* $X$ is a function from $T$ to $S$ such that $f(X) \in \mathscr{Y}$ for each $f$ in $\mathscr{X}$. A probability $P^{\mathscr{X}}$ is induced on $\mathscr{X}$ by

$$P^{\mathscr{X}} f = P[f(X)] \quad \text{each } f \text{ in } \mathscr{X}.$$

The distribution of $X$ is defined to be $P^{\mathscr{X}}$, also denoted by $P^X$.

If $S$ is the real line, and $\mathscr{X}$ is the smallest probability space including finite intervals, from 2.2, $P^X$ is determined by the values it gives finite intervals $P\{a < X \leqq b\} = G(b) - G(a)$. The distribution function $G$ is right continuous and uniquely determined up to an additive constant. If $\sup_a P(a < X \leqq 0\} < \infty$, set $G(b) = \sup_a P\{a < X \leqq b\}$. If $P^X$ is unitary, it will follow that $\lim_{a \to -\infty} G(a) = 0$, $\lim_{a \to \infty} G(a) = 1$.

## 2.4. Probable Bets

Let $\mathscr{X}$ be a linear space of bets: $X, Y \in \mathscr{X} \Rightarrow aX + bY \in \mathscr{X}$ for $a, b$ real.

Let $\mathscr{P}$, the *probable set*, be a cone of bets: $X, Y \in \mathscr{P} \Rightarrow aX + bY \in \mathscr{P}$ for $a, b \geqq 0$.

A *generalized probability* $P$ for $\mathscr{P}$ is a linear functional on $\mathscr{P}$ (a real valued function on $\mathscr{P}$ with $P(aX + bY) = aPX + bPY$) such that $PX \geqq 0$ for $X$ in $\mathscr{P}$, $PX > 0$ some $X$ in $\mathscr{P}$. *For sections 2.4 and 2.5, $P$ will be referred to as a probability.*

**Theorem.** *If $\mathscr{P} \neq \mathscr{X}$ and $\mathscr{P}$ contains an internal point (a point $X_0$ such that for every $X$ in $\mathscr{X}$, $X + kX_0 \in \mathscr{P}$ for some $k$), then a probability $P$ exists for $\mathscr{P}$.*

PROOF. [*Following Dunford and Schwartz (1964), p. 412.*]

Let $N = \mathscr{P} \cap (-\mathscr{P})$ be the neutral set of bets, bets $X$ such that both $X$ and $-X$ are probable. If $\{\mathscr{P}_\alpha\}$ is a chain of probable sets with neutral sets $N$, then $\cup \mathscr{P}_\alpha$ is a probable set with neutral set $N$. From Zorn's lemma, there is a maximal probable set $\mathscr{P}_0$ containing $\mathscr{P}$ and having neutral set $N$. Then $\mathscr{X} = \mathscr{P}_0 \cup (-\mathscr{P}_0)$, for if $X \notin \mathscr{P}_0$, $X \notin -\mathscr{P}_0$ the set $\mathscr{P}_0(X) = \{\alpha X + Y, \alpha \geqq 0, Y \in \mathscr{P}_0\}$ is a probable set with neutral set $N$ and $\mathscr{P}_0(X)$ strictly includes $\mathscr{P}_0$.

The internal point $X_0$ does not lie in $-\mathscr{P}$ for $X = X + kX_0 + k(-X_0)$ would lie in $\mathscr{P}$ for each $X$ contradicting $\mathscr{P} \neq \mathscr{X}$. Also $X_0$ is an internal point for $\mathscr{P}_0$. Define $PX = \sup\{\alpha \mid X - \alpha X_0 \in \mathscr{P}_0\}$; then $-\infty < PX < \infty$ since $X + kX_0 \in \mathscr{P}_0$ and $-X + k'X_0 \in \mathscr{P}_0$ some $k, k'$; $P$ is a linear functional because $X - \alpha X_0 \in \mathscr{P}_0$ or $-\mathscr{P}_0$ for every $X, \alpha$; for $X \in \mathscr{P} \subset \mathscr{P}_0$, $PX \geqq 0$; and $PX_0 = 1$. Thus $P$ is a probability for $\mathscr{P}$. ☐

It is necessary that $\mathscr{P} \neq \mathscr{X}$, for otherwise we cannot separate probable bets from others, and it is necessary to assume an internal point so that one of the probable bets will be comparable to all possible bets.

It is usual to take the bets as real values received according to whichever state of nature occurs, but it is not necessary to do so. See Ramsey (1926) and Savage (1954). The space of bets and probable set may be constructed from a preference ordering among a set of mixed actions as follows. Let $\mathscr{A}$ be an arbitrary set (not necessarily countable) of actions $a_1, a_2, \ldots$ ; let $\mathscr{A}^*$ be the mixed actions $\sum_{i=1}^n p_i a_i$ (perhaps constructed by generating new actions by taking action $a_i$ with chance $p_i$) where $p_i \geqq 0$, $\sum p_i = 1$; and let $\geqq$ be a preference between mixed actions such that for $0 \leqq \alpha \leqq 1$, $a \geqq \alpha a + (1 - \alpha)a \geqq a$ and $a \geqq b$, $c \geqq d \Rightarrow \alpha a + (1 - \alpha)c \geqq \alpha b + (1 - \alpha)d$. Construct the space $\mathscr{X}$ of bets $\sum_{i=1}^n x_i a_i$ where $\sum x_i = 0$, and define the probable set $\mathscr{P}$ to consist of bets $\lambda(a - b)$ where $\lambda \geqq 0$ and $a \geqq b$, $a, b \in \mathscr{A}^*$. A probability $P$ for $\mathscr{P}$ will now satisfy $Pa \geqq Pb$ whenever $a \geqq b$ and $Pa > Pb$ for at least one pair $a \geqq b$. The condition that an internal point exists is equivalent to assuming a pair $a_0 \geqq b_0$ such that for each $a, b, \alpha a + (1 - \alpha)a_0 \geqq \alpha b + (1 - \alpha)b_0$ some $\alpha$, $0 \leqq \alpha \leqq 1$.

On the other hand there is no harm in assuming that bets are real valued functions. Assume that $\mathscr{P} \neq \mathscr{X}$, and that an internal point exists. Then there exists a basis of $\mathscr{X}$, $\{X_\alpha\}$, $X_\alpha \in \mathscr{P}$ such that each $X$ in $\mathscr{X}$ is represented uniquely by $\sum c_\alpha X_\alpha$ where only a finite number of $c_\alpha$ are non-zero, and so $X$ corresponds to the real valued function $f$, $f(\alpha) = c_\alpha$. Note that $X \in \mathscr{P}$ whenever $f \geqq 0$.

## 2.5. Comparative Probability

In comparative probability, all pairs of events in $\mathscr{F}$ are compared by the relation $\leq$, "is no more probable than":

(i) $\phi \leq A$ for $A \in \mathscr{F}$,
(ii) $S \leq \phi$ is not true,
(iii) $A \leq B$, $C \leq D$ implies $A + C \leq B + D$ if $AC = BD = 0$.

The statements $A \leq B$ may be interpreted as offering the bet: pay 1 unit if $A$ occurs to receive 1 unit if $B$ occurs. The family of bets $\sum_{i=1}^{n} \alpha_i(B_i - A_i)$ for $B_i$, $A_i$ in $\mathscr{F}$ forms a betting space; suppose the statements $A_i \leq B_i$ are construed as accepting all bets $\sum_{i=1}^{n} \alpha_i(B_i - A_i)$ for which $\alpha_i \geq 0$. It may be possible to make a book against the bets $\{A_i \leq B_i\}$: find a linear combination $\sum_{i=1}^{n} \alpha_i(B_i - A_i)$ which is negative. Otherwise, for $S$ finite, the set of combinations of $\leq$ bets is probable, and there exists a probability $P$ on $\mathscr{F}$ such that $A \leq B$ implies $P(A) \leq P(B)$. An example of such a "beatable" comparative probability is given by Kraft, et al. (1959) for a set of 5 elements. See also Scott (1964) who connects "unbeatability" with the existence of a conforming numerical probability, and Fine (1973) for a general discussion, and for continuity axioms.

The above axioms of comparative probability are unsatisfactory because they may not generate a probable set of bets. One solution to the problem is to prohibit negative combinations, which is just equivalent to requiring that a certain subset of a betting space, generated by pairs of events, is probable. An alternative approach followed by Koopman (1940) and Savage (1954) supposes that $S$ may be partitioned into sets of arbitrarily small probability; Koopman requires that for each $n$ there exist a partition into $n$ events of equal probability. Since all pairs of events are comparable, each event has a precise numerical probability determined by comparison with events in increasingly fine partitions, and this numerical probability satisfies the usual finitely additive axioms with $P(S) = 1$.

## 2.6. Problems

Exercises (E), are supposed to be easier than problems (P).
Probability is used in the sense of Section 2.4.

E1. Let $\mathscr{X} = \mathbb{R}^2$, define $\mathscr{P} = \{x, y \mid y + x \geq 0, \ y + \frac{1}{2}x \geq 0\}$. Show that $\mathscr{P}$ is a probable set, and find all probabilities $P$ such that $P(X) \geq 0$ for $X$ in $\mathscr{P}$.

P1. A bookie offers the following odds for various teams to win a basketball pennant.

Knicks: 6/1      Bullets: 2/1      Braves: 2/1      Celtics: 1/1

Odds of 6/1 means that he receives \$1 if the Knicks lose and pays \$6 if the Knicks win. Consider the space of bets $\mathscr{X}\{(x_1, x_2, x_3, x_4)\}$ in which the bookie receives

$x_i$ if the $i$th team wins. Show that any probable set including the specified bets will include all bets.

E2. Let $\mathscr{X} = \mathbb{R}^k$. Let $P$ be a probability on $\mathscr{X}$ with $PX \geqq 0$ for $X \in \mathscr{X}$, $X \geqq 0$. Show that there exist $p_1, \ldots, p_k, p_i \geqq 0$ such that $P(X) = \sum_{i=1}^{k} p_i X_i$, where $X_i$ denotes the $i$th co-ordinate of $X$.

E3. Let $\mathscr{X}$ consist of linear combinations of bets $\{s \mid a < s \leqq b\}$, $a < b$. Let $\mathscr{P}$ consist of non-negative combinations of bets $(a, a + 2\delta] - (a - \delta, a]$. Find a probability on $\mathscr{X}$ which is positive for all nonzero bets in $\mathscr{P}$.

E4. Let $\mathscr{X}$ be the real sequences, and let $\mathscr{P}$ consist of sequences $X$ with $\varliminf \sum_{i=1}^{n} X_i \geqq 0$. Show that if a probability $P$ on $(\mathscr{X}, \mathscr{P})$ is such that $X_0 = (1, 1, 1, \ldots)$ has $P(X_0) = 1$, the positive sequence $X = (1, \frac{1}{2}, \frac{1}{3}, \ldots, 1/n, \ldots)$ has $P(X) = 0$.

E5. Let $\mathscr{X}$ be the real sequences, $X = (X_1, X_2, \ldots)$, with finitely many non-zero elements, and let $\mathscr{P} = \{X \mid \text{for some } i, X_i > 0, X_{i+1} \geqq 0, X_{i+2} \geqq 0, \ldots\} \cup \{0\}$. If $P$ is a probability on $(\mathscr{X}, \mathscr{P})$, show that $P\{i\} = 0$ or $\infty$ except for one $\{i\}$, where $\{i\}$ is the bet equal to 1 at $i$ and zero elsewhere.

P2. Let $S$ be the real line, $\mathscr{F}$ be the ring of unions of half open intervals $(a < s \leqq b)$, where $-\infty \leqq a < b \leqq \infty$. Define $P((a, b]) = F(b) - F(a)$ where $F$ is a non-decreasing right continuous function. Show that $P$ is a probability on $\mathscr{F}$, in the sense of Section 2.1.

E6. Let $\mathscr{X}$ be $k$-dimensional euclidean space, and let the probable set $\mathscr{P}$ include all bets $X = (X_1, \ldots, X_k)$ such that $X_i \geqq 0$, $1 \leqq i \leqq k$. Show that if $\mathscr{P}$ is not neutral, $\mathscr{P}$ includes no bet which is uniformly negative.

E7. A bookmaker offers a number of bets $X$, $Y$, ... in $k$-dimensional euclidean space; the bet $X = (X_1, \ldots, X_k)$ means he receives $X_i$ if $i$ occurs. Show that there is some mixture of the bets on which he always receives a negative pay off, or else there is a probability $P$ which is non-negative for all bets.

E8. Let $\mathscr{X}$ be the set of real-valued sequences $X = (x_1, x_2, \ldots, x_n, \ldots)$, let $(p_1, p_2, \ldots, p_n, \ldots)$ be a fixed sequence, $p_i \geqq 0$, and let $\mathscr{P}$ be the sequences $X$ with $\varliminf \sum_{i=1}^{n} p_i X_i \geqq 0$. Show that $\mathscr{P}$ is a probable set and specify the probability which gives value 1 to $(1, 1, \ldots, 1, \ldots)$, and the probability which gives value 1 to $(1, 0, \ldots, 0)$. Show that the first probability is continuous if and only if $\sum p_i < \infty$, and the second probability is bounded if and only if $\sum p_i < \infty$.

E9. Replace the third axiom of comparative probability by

(iii)' if $\sum_{i=1}^{n} (A_i - B_i) = \sum_{i=1}^{n} (A_i' - B_i')$, and all $A_i \leqq B_i$, then at least one $A_i' \leqq B_i'$.

Then the set of bets $\sum_{i=1}^{n} \alpha_i (B_i - A_i)$ where $A_i \leqq B_i$, $\alpha_i \geqq 0$ is a probable set in the space of real-valued function on $S$.

P3. The axioms of comparative probability are satisfied by subsets of $S = (1, 2, 3, 4, 5)$ with $\varnothing < 2 < 3 < 4 < 23 < 24 < 1 < 12 < 34 < 5 < 234 < 13 < 14 < 25 < 123 < 35 < 124$, the remaining sets being ordered by complements. Show that no numerical probability conforms to the order. [Kraft, et al., 1959.]

E10. Add a fineness axiom to the axioms of comparative probability:

(iv) for each $n$, there exists $\{A_i\}$ with $\sum_{i=1}^{n} A_i = S$, $A_i A_j = 0$, $A_i \leqq A_j$ each $i, j$.

Then there is a unique probability with $A \leq B \Rightarrow P(A) \geq P(B)$, $A, B \in \mathscr{F}$. [*Koopman*, 1940.]

P4. Let a finitely additive probability $P$ be defined on the plane so that $P(|x| + |y| > a) = 0$ for $a > 0$, $P[x < 0] = P[x = 0] = P[x > 0] = \frac{1}{3}$, $P[y < 0] = P[y = 0] = P[y > 0] = \frac{1}{3}$, events determined by $x$ are independent of $y$, and (*A*) $P[x + y = 0, x > 0, y < 0] = P[x + y = 0, x < 0, y > 0] = \frac{1}{9}$. These conditions determine $P$ uniquely. Show that a different $P$ is determined if (*A*) is replaced by $P[x + y < 0, x > 0, y < 0] = P[x + y < 0, x < 0, y > 0] = \frac{1}{9}$, demonstrating that the distribution of $x + y$ is not determined from the distributions of $x$ and $y$ when $x$ and $y$ events are independent.

P5. Let $X$ be a random variable from $U, Z$ to $S, \mathscr{X}$ where $S$ is the real line and $\mathscr{X}$ includes all finite intervals. Show that the $P$-completion of $Z$ includes $X$ if $\sum_{k=0}^{\infty} P^X \{ |s| \leq k \} < \infty$.

## 2.7. References

De Finetti, B. (1970), *Theory of Probability*, Vol. 1. John Wiley: London.
De Finetti, B. (1972), *Theory of Probability*, Vol. 2. John Wiley: London.
Dunford, N. and Schwartz, J. T. (1964), *Linear Operators*, Part I. John Wiley: New York.
Fine, T. (1973), *Theories of Probability, an Examination of Foundations*. New York: Academic Press.
Jeffreys, H. (1939), *The Theory of Probability*. London: Oxford University Press.
Keynes, J. M. (1921), *A Treatise on Probability*. New York: Harper.
Kolmogorov, A. N. (1950), *Foundations of the Theory of Probability*. New York: Chelsea.
Koopman, B. O. (1940), The bases of probability, *Bull. Am. Math. Soc.* **46**, 763–774.
Kraft, C., Pratt, J. and Seidenberg, A. (1959), Intuitive probability on finite sets, *Ann. Math. Statist.* **30**, 408–419.
Loomis, L. H. (1953), *An Introduction to Abstract Harmonic Analysis*. Princeton: Van Nostrand.
Renyi, A. (1970), *Probability Theory*. New York: American Elsevier.
Ramsey, F. P. (1926), Truth and probability, reprinted in H. E. Kyburg and H. E. Smokler (eds.), *Studies in Subjective Probability*. New York: John Wiley, 1964, pp. 61–92.
Savage, L. J. (1954), *The Foundations of Statistics*. New York: John Wiley.
Scott, D. (1964), Measurement structures and linear inequalities, *J. Math. Psych.* **1**, 233–247.

# CHAPTER 3

# Conditional Probability

## 3.0. Introduction

Kolmogorov's exquisite formalization of conditional probability in the unitary case (1933) does not readily generalize to non-unitary probabilities. Stone and Dawid (1972) show one type of difficulty with their marginalization paradoxes for improper priors.

Consider the case of the uniform distribution over pairs of positive integers $\{i, j\}$. The desired conditional distribution of $\{i, j\}$ given $j = j_0$ is uniform over $i$. Following Kolmogorov, the conditional distribution given $j$ should combine with the marginal distribution to return the joint distribution:

$$p(i, j) = p(i \mid j_0) p(j_0).$$

But the event $[\{i, j\}, j = j_0]$ is not given a probability so the marginal probabilities $p(j_0)$ are not determined by $p(i, j)$, $1 \leq i, j \leq \infty$. Correspondingly the uniform distribution over $\{i, j\}$ given $j = j_0$ is equally well represented by $p(i \mid j_0) = k(j_0)$ for any $k(j_0)$. Thus although these conditional distributions are determined by the joint distribution, the marginal distribution is not. (This is the explanation of the marginalization paradoxes of Stone and Dawid.)

It is assumed therefore that the joint distribution, the conditional distribution, and the marginal distribution are specified separately to follow the axioms of conditional probability. In particular the probabilities of the $\{i, j\}$ and of $[\{i, j\}, j = j_0]$ are separately specified. We are declaring that $\{i, j\}$ has the same probability as $\{i', j'\}$, and *in addition* that $[\{i, j\}, j = j_0]$ has the same probability as $[\{i, j\}, j = j_0']$.

23

## 3.1. Axioms of Conditional Probability

Let $\mathscr{X}$, $\mathscr{Y}$, $\mathscr{Z}$, ... be probability spaces of functions on $S$. The conditional probability on $\mathscr{X}$ given $\mathscr{Y}$ is a function $P$ from $\mathscr{X}$ to $\mathscr{Y}$ that is

LINEAR: $P(Y_1 X_1 + Y_2 X_2) = Y_1 P X_1 + Y_2 P X_2$ for $X_i \in \mathscr{X}$, $Y_i \in \mathscr{Y}$, $Y_i X_i \in \mathscr{X}$
NON-NEGATIVE: $P|X| \geq 0$
CONTINUOUS: $PX_n \to PX$ for $|X_n| \leq X_0 \in \mathscr{X}$, $X_n \to X$
INVARIANT: If $Y \in \mathscr{X} \cap \mathscr{Y}$, $PY = Y$

A family of conditional probabilities is assumed to satisfy the

PRODUCT RULE: If $P_{\mathscr{Y}}^{\mathscr{X}}$, $P_{\mathscr{Z}}^{\mathscr{Y}}$, $P_{\mathscr{Z}}^{\mathscr{X}}$ denote conditional probabilities from $\mathscr{X}$ to $\mathscr{Y}$, $\mathscr{Y}$ to $\mathscr{Z}$ and $\mathscr{X}$ to $\mathscr{Z}$ respectively,

$$P_{\mathscr{Z}}^{\mathscr{X}} = P_{\mathscr{Z}}^{\mathscr{Y}} P_{\mathscr{Y}}^{\mathscr{X}}.$$

The conditional probability $P$ is determined as a probability on $\mathscr{X}$ given the results of an experiment which determines the values of all functions in $\mathscr{Y}$. Each result of the experiment will give rise to possibly different values of functions in $\mathscr{Y}$, and possibly different probabilities. The conditional probability $P$ determines these different probabilities for all possible results of the experiment. If $PX \in \mathscr{X}$, then $PX$ may be interpreted as a bet equivalent to $X$ that has known value after the experiment is performed.

The above axioms generalize the axioms of probability. Let $\mathscr{X}_1$ denote the probability space of constant functions on $S$, and let $\mathbf{1}$ denote the constant function. Then $P$ is a probability on $\mathscr{X}$ if and only if $P_{\mathscr{X}_1}^{\mathscr{X}} : X \to (PX)\mathbf{1}$ is a conditional probability on $\mathscr{X}$ given $\mathscr{X}_1$. Indeed $P_{\mathscr{X}_1}^{\mathscr{X}} = P_{\mathscr{X}_1}^{\mathscr{Y}} P_{\mathscr{Y}}^{\mathscr{X}}$ implies that $P_{\mathscr{Y}}^{\mathscr{X}}$ is determined almost uniquely by $P_{\mathscr{X}_1}^{\mathscr{X}}$ and $P_{\mathscr{X}_1}^{\mathscr{Y}}$. [Suppose $P_{\mathscr{Y}}^{\mathscr{X}} X$ could have values $Y_1$ or $Y_2$; then $P_{\mathscr{X}_1}^{\mathscr{Y}}[Y(Y_1 - Y_2)] = 0$ all $Y \in \mathscr{Y}$, so $P_{\mathscr{X}_1}^{\mathscr{Y}}|Y_1 - Y_2| = 0$.] Kolmogorov (1933) defines conditional probability in terms of probability; under certain regularity conditions, there exists a "conditional probability" that satisfies the above axioms except on a subset of $S$ of probability zero. Here we are following the more traditional scheme of axiomatizing conditional probability rather than defining it in terms of probability.

EXAMPLE 1: *Toss a penny twice.* Let $\mathscr{X}$ be the bets $\{X_{HH}, X_{HT}, X_{TH}, X_{TT}\}$ where $X_{HH}$ means the amount received if two heads occur, and similarly for the other three results. The result of the first toss of the experiment, heads or tails, determines the values of all bets in $\mathscr{Y} = \{\mathbf{X} | X_{HH} = X_{HT}, X_{TH} = X_{TT}\}$, bets of form $(X_H, X_H, X_T, X_T)$. Assuming that tails and heads have probability 1/2 given the results of the first toss, the conditional probability of $X = (X_{HH}, X_{HT}, X_{TH}, X_{TT})$ is $P_{\mathscr{Y}} X = (X_H, X_H, X_T, X_T)$ where $X_H = \frac{1}{2}(X_{HH} + X_{HT})$ and $X_T = \frac{1}{2}(X_{TH} + X_{TT})$. Here $P_{\mathscr{Y}} X$ is a bet equivalent to $X$ that has known value, either $X_H$ or $X_T$, once the first toss is known.

Suppose that head on the first toss has probability $p$, and tail has probability $(1 - p)$. Then

$$P^{\mathscr{X}}_{\mathscr{X}_1} X = P^{\mathscr{Y}}_{\mathscr{X}_1} P^{\mathscr{X}}_{\mathscr{Y}} X$$
$$= P^{\mathscr{Y}}_{\mathscr{X}_1} [X_H, X_H, X_T, X_T]$$
$$= p X_H + (1 - p) X_T$$
$$= \tfrac{1}{2} p X_{HH} + \tfrac{1}{2} p X_{HT} + \tfrac{1}{2}(1 - p) X_{TH} + \tfrac{1}{2}(1 - p) X_{TT}.$$

The probability on $\mathscr{X}$ corresponds to giving probability $\frac{1}{2}p$, $\frac{1}{2}p$, $\frac{1}{2}(1 - p)$, $\frac{1}{2}(1 - p)$ to the four outcomes $HH$, $HT$, $TH$, $TT$. These probabilities have been developed from conditional probability using the product rule, but in the finite case we could just as well define conditional probability in terms of probability; a separate axiomatization of conditional probability is necessary only in the infinite case.

EXAMPLE 2: *Uniform distribution on the square.* Let $\mathscr{X}$ denote the smallest probability space of functions including the continuous functions on the square; let $X(u, v)$ denote the value of $X$ at the point $(u, v)$, $0 \leq u$, $v \leq 1$. Let $\mathscr{Y}$ denote the set of functions in $\mathscr{X}$ depending only on $u$—$Y(u, v) = Y(u, 1)$ for all $v$.
Define $P^{\mathscr{X}}_{\mathscr{Y}} X = \int X(u, v) dv$.
Define $P^{\mathscr{Y}}_{\mathscr{X}_1} Y = \int Y(u, v) du$.
Here $P^{\mathscr{X}}_{\mathscr{Y}} X$ is a bet equivalent to $X$ that is a function of $u$ alone. From the product axiom, $P^{\mathscr{X}}_{\mathscr{X}_1} X = \int (\int X(u, v) dv) du = \int \int X(u, v) du dv$. Thus the probability on $\mathscr{X}$ is just Lebesgue measure, in which the probability of a set is its area. Beginning with $P^{\mathscr{X}}_{\mathscr{X}_1}$, it is possible to construct the conditional probability $P^{\mathscr{X}}_{\mathscr{Y}}$ to satisfy the product axiom almost uniquely—any other solution $Q^{\mathscr{X}}_{\mathscr{Y}}$ satisfies $P |(P^{\mathscr{X}}_{\mathscr{Y}} - Q^{\mathscr{X}}_{\mathscr{Y}}) X| = 0$. Tjur (1974) assures uniqueness by requiring continuity of the conditional probability, but then, establishing existence is sometimes formidable.

EXAMPLE 3: *Uniform distribution on the integers.* Let $\mathscr{X}$ be the space of sequences $\{x_1, x_2, \ldots, x_n, \ldots\}$ with $\sum |x_i| < \infty$, let $\mathscr{Y}$ be the space of sequences $\{\alpha, \beta, \alpha, \beta, \ldots\}$.
Define $P^{\mathscr{X}}_{\mathscr{Y}} X = \{\alpha(X), \beta(X), \alpha(X), \beta(X), \ldots\}$ where $\alpha(X) = 2 \sum_{i=1}^{\infty} x_{2i-1}$, $\beta(X) = 2 \sum_{i=1}^{\infty} x_{2i}$.
Define $P^{\mathscr{Y}}_{\mathscr{X}_1} Y = \tfrac{1}{2}\alpha + \tfrac{1}{2}\beta$.
Then $P^{\mathscr{X}}_{\mathscr{X}_1} X = P^{\mathscr{Y}}_{\mathscr{X}_1} P^{\mathscr{X}}_{\mathscr{Y}} X = \sum_{i=1}^{\infty} x_i$.
Here $P^{\mathscr{X}}_{\mathscr{Y}}$ is a uniform distribution on the evens and the odds, according as $Y = (0, 1, 0, 1, 0, 1, \ldots)$ is one or zero. The conditional distribution is not unitary. The distribution on $\mathscr{Y}$ gives probability $1/2$ to the evens, probability $1/2$ to the odds; this distribution is unitary. The distribution on $\mathscr{X}$ is uniform over the integers; it is non-unitary.
Note that $P^{\mathscr{X}}_{\mathscr{Y}}$ is not determined by saying that it is uniform given evens and uniform given odds; probabilities given even must be compared explicitly

with probabilities given odds (such comparisons are implicit when probabilities are unitary, since 1 has the same probability under all conditions).

Let the conditioning family $\mathcal{Z}$ be the sets of sequences $X$ with $x_{2i} = x_{2i-1}$, $i = 1, 2, \ldots,$. Define

$$P_{\mathcal{Z}}^{\mathcal{X}} X = \{\tfrac{1}{2}x_1 + \tfrac{1}{2}x_2, \tfrac{1}{2}x_1 + \tfrac{1}{2}x_2, \tfrac{1}{2}x_3 + \tfrac{1}{2}x_4, \tfrac{1}{2}x_3 + \tfrac{1}{2}x_4, \ldots\}$$
$$P_{\mathcal{X}_1}^{\mathcal{Z}} Z = 2\sum Z_i.$$

Then again $P_{\mathcal{X}_1}^{\mathcal{X}} = \sum x_i$.

Here the conditional probability is uniform given 1 or 2, or given 3 or 4, or given 5 or 6, ... The conditional probability is unitary; the probability on $\mathcal{Z}$ is non-unitary. In this case, some elements of $\mathcal{Z}$ lie in $\mathcal{X}$ and for these $P_{\mathcal{Z}}^{\mathcal{X}} Z = Z$, satisfying invariance.

EXAMPLE 4: *Uniform distribution in the plane.* Let $\mathcal{X}$ be the smallest probability space which includes continuous functions zero outside some square in the plane, $-\infty \leq u, v \leq \infty$. Let $\mathcal{Y}$ be the probability space of such functions depending only on $u$.

Define $\qquad\qquad P_{\mathcal{Y}}^{\mathcal{X}} X = \int X(u, v) dv$
$$P_{\mathcal{X}_1}^{\mathcal{Y}} Y = \int Y(u, v) du.$$
Then $\qquad\qquad P_{\mathcal{X}_1}^{\mathcal{X}} X = \int (\int X(u, v) dv) du = \int X(u, v) du dv,$

corresponding to the uniform distribution on the plane.

Note that the conditional distribution is not determined by requiring that the distribution be uniform given each $u$; since the uniform distribution is non-unitary, it is possible to have a conditional distribution which is uniform given each $u$, a distribution on $u$ which is not uniform, a distribution over the whole plane which is uniform. The marginal distribution on $\mathcal{Y}$ is *not* determined by the distribution on $\mathcal{X}$; the conditional distribution is determined only up to an arbitrary weighting factor depending on $u$. Given the distribution on $\mathcal{X}$ and the distribution on $\mathcal{Y}$, the product axiom determines $P_{\mathcal{Y}}^{\mathcal{X}}$ almost uniquely.

## 3.2. Product Probabilities

For arbitrary $S$ and $T$ define the function subscript $S$, denoted by $_S$, by $_S(s)(t) = (s, t)$. Thus $_S$ is a function from $S$ to the space of functions from $T$ to $S \times T$. Define $_T$ similarly.

**Fubini's Theorem.** *Let $P$ be a probability on $\mathcal{X}$ on $S$, let $Q$ be a probability on $\mathcal{Y}$ on $T$, let $X \times Y$ be the function on $S \times T$: $(s, t) \to X(s)Y(t)$, and let $\mathcal{X} \times \mathcal{Y}$ be the smallest probability space including all $X \times Y$.*

*Then $P \times QW = PQW_S = QPW_T$ is the unique probability on $\mathcal{X} \times \mathcal{Y}$ such that $P \times QX \times Y = PX\,QY$. (Note that $QW_S$ is the function $s \to QW_S(s)$.)*

PROOF. Let $\mathcal{W}_0$ be the set of functions $W$ in $\mathcal{X} \times \mathcal{Y}$ such that $W_s(s) \in \mathcal{Y}$ for each $s \in \mathcal{X}$. Then $\mathcal{W}_0$ includes all functions $X \times Y$, and is a probability space, so $\mathcal{W}_0 = \mathcal{X} \times \mathcal{Y}$.

Again let $\mathcal{W}_0$ be the set of functions $W$ in $\mathcal{X} \times \mathcal{Y}$ such that $QW_s \in \mathcal{X}$. Then $\mathcal{W}_0$ is linear, includes all functions $X \times Y$ and is continuous, but it is not straightforward to show that $W \in \mathcal{W}_0 \Rightarrow |W|, W \wedge 1 \in \mathcal{W}_0$. Let $\mathcal{A}(\mathcal{X})$ be the set of functions $\sum_{i=1}^{n} a_i X_i$, let $\mathcal{A}(\mathcal{Y})$ be the set of functions $\sum_{i=1}^{n} a_i Y_i$ and let $\mathcal{A}(\mathcal{X}, \mathcal{Y})$ be the set of functions $\sum a_i X_i \times Y_i$ where $X_i$ and $Y_i$ are $0 - 1$ functions on $S$ and $T$. For each $X \in \mathcal{X}$ there is a sequence $X_n \to X$, $|X_n| \leq X$, $X_n \in \mathcal{A}(\mathcal{X})$. Thus if $X^i \in \mathcal{X}$, $Y^i \in \mathcal{Y}$, then $|X^1 \times Y^1 + X^2 \times Y^2| = \lim X_n^1 \times Y_n^1 + X_n^2 \times Y_n^2$, where $X_n^1 \times Y_n^1 + X_n^2 \times Y_n^2 \in \mathcal{A}(\mathcal{X}, \mathcal{Y})$ and is bounded by $|X^1| \vee |X^2| \times |Y^1| \vee |Y^2|$. Since $\mathcal{A}(\mathcal{X}, \mathcal{Y}) \subset \mathcal{W}_0$, and $\mathcal{W}_0$ is continuous, $|X^1 \times Y^1 + X^2 \times Y^2|$ lies in $\mathcal{W}_0$. By a similar argument, any finite sequence of operations involving linear combinations or absolute values or $\wedge 1$ on the functions $X \times Y$ will yield a function in $\mathcal{W}_0$, so that the prespace including all $X \times Y$ is included in $\mathcal{W}_0$. Since $\mathcal{W}_0$ is continuous, by Lemma 2.2, $\mathcal{W}_0 = \mathcal{X} \times \mathcal{Y}$.

Define $P \times QW = PQW_s$; note that

$$PQX \times Y_s = P(XQY) = PX\,QY.$$

It is easy to show that $P \times Q$ is a probability. For example, continuity requires $W_n \to W$, $|W_n| \leq W_0 \Rightarrow P \times Q_n W \to P \times QW$. For each $s$, $W_{ns}(s) \to W_s(s)$, $|W_{ns}(s)| \leq W_{0s}(s)$ so $QW_{ns}(s) \to QW_s(s)$ and $|QW_{ns}| \leq |QW_s|$. Therefore $PQW_{ns} \to PQW_s$ as required.

Also $P \times Q$ is the only probability on $\mathcal{X} \times \mathcal{Y}$ such that $P \times QX \times Y = PX\,QY$; for any $W$ in $\mathcal{X} \times \mathcal{Y}$ may be approximated by a sequence of functions of form $\sum_{n=1}^{N} a_n X_n \times Y_n$. By symmetry $P \times QW = QPW_T$, and the theorem is proved.                                                                                $\square$

## 3.3. Quotient Probabilities

Let $\mathcal{X}$, $\mathcal{Z}$ be probability spaces on $S$ and let $Y$ be a random variable from $S$, $\mathcal{Z}$ to $T$, $\mathcal{Y}$. The *conditional probability of* $\mathcal{X}$ *given* $Y$ is defined by $(P_Y^{\mathcal{X}} X)(Y) = P_{Y^{-1}(\mathcal{Y})}^{\mathcal{X}} X$. Thus for each $X$, $P_Y^{\mathcal{X}} X$ is a function in $\mathcal{Y}$, such that $P_Y^{\mathcal{X}} X(t)$ is a probability on $\mathcal{X}$ for each $t$. The notation $P(X \mid Y)$ means $(P_Y^{\mathcal{X}} X)(Y)$, a function in $Y^{-1}(\mathcal{Y})$.

Suppose that $X$, $Y$ and $X \times Y$ are random variables on $U$, $\mathcal{Z}$ to $S$, $\mathcal{X}$, $T$, $\mathcal{Y}$ and $S \times T$, $\mathcal{X} \times \mathcal{Y}$, and that there exists a conditional probability on $(X \times Y)^{-1}\mathcal{X} \times \mathcal{Y}$ given $Y^{-1}(\mathcal{Y})$. The *conditional probability of* $X$, $Y$ *given* $Y$ is $(P_Y^{X,Y} W)(Y) = P_{Y^{-1}(\mathcal{Y})}^{(X \times Y)^{-1}\mathcal{X} \times \mathcal{Y}} W(X, Y)$, $W \in \mathcal{X} \times \mathcal{Y}$. Thus for each $W$ in $\mathcal{X} \times \mathcal{Y}$, $P_Y^{X,Y} W$ is a function in $\mathcal{Y}$, such that $(P_Y^{X,Y} W)(t)$ is a probability on $\mathcal{X} \times \mathcal{Y}$ for each $t$. The notation $P[W(X, Y) \mid Y]$ means $(P_Y^{X,Y} W)(Y)$ and is useful for indicating that $X$ is summed over while $Y$ is held fixed. The product rule becomes $P^{X,Y} = P^Y P_Y^{X,Y}$.

A *quotient probability* $P_Y^X$ is a probability on $\mathcal{X}$ for each $t$ such that $g P_Y^X f \in \mathcal{Y}$

for each $g \in \mathscr{Y}, f \in \mathscr{X}$. A *conditional probability* $P_Y^{X,Y}$ is defined from a quotient probability by $P_Y^{X,Y} fg = g P_Y^X f$ each $g \in \mathscr{Y}, f \in \mathscr{X}$; this equation determines $P_Y^{X,Y}$ on $\mathscr{X} \times \mathscr{Y}$. A quotient probability is not a conditional probability because $P_Y^X f$ may not lie in $\mathscr{Y}$ for each $f$ in $\mathscr{X}$; it is convenient to use quotient probabilities to generate conditional probabilities because it is necessary to specify probabilities on $\mathscr{X}$ for each $t$, rather than on $\mathscr{X} \times \mathscr{Y}$ for each $t$. As before $P[f(X)|Y]$ means $(P_Y^X f)(Y)$.

The random variables $X$ and $Y$ are *independent* if $P^{X,Y} fg = P^X f P^Y g$ for $f \in \mathscr{X}$, $g \in \mathscr{Y}$ or equivalently, given that $P_Y^X$ is defined, if $P_Y^X = P^X$. Similarly, the random variables $\{X_\alpha\}$ are *independent* if for any finite subset $X_1, \ldots, X_n$

$$P^{X_1, \ldots, X_n} \prod f_i = \prod P^{X_i} f_i, f_i \in \mathscr{X}_i.$$

The random variables $X$ and $Y$ are *conditionally independent given* $Z$ if $P_Z^{X,Y} fg = P_Z^X f P_Z^Y g, f \in \mathscr{X}, g \in \mathscr{Y}$, that is, if $P_{Y,Z}^X = P_Z^X$.

EXAMPLE. Let $S$ be the set of positive integer pairs $(i, j)$, $i \geq j$. Let $\mathscr{X}$ be the probability space of functions $X, \sum_{i \geq j} |X(i, j)| < \infty$. Let $\mathscr{Z}$ be the probability space of real valued functions on $S$. Let $Y$ be the function $Y(i, j) = j$ from $S, \mathscr{Z}$ to $T, \mathscr{Y}$ where $T$ denotes the positive integers and $Y$ consists of functions $g$ where $\sum |g(i)| < \infty$.

A *conditional probability* $P_Y^{\mathscr{X}}$ is

$$(P_Y^{\mathscr{X}} X)(j) = \sum_{i \geq j} X(i, j).$$

For each $j$, $(P_Y^{\mathscr{X}} X)(j)$ defines a probability on $\mathscr{X}$. For each $X$, $P_Y^{\mathscr{X}} X$ is a function in $\mathscr{Y}$. The function $(P_Y^{\mathscr{X}} X)(Y)$ has value at $(i, j)$: $P_Y^{\mathscr{X}} X Y(i, j) = (P_Y^{\mathscr{X}} X)(j) = \sum_{i \geq j} X(i, j)$ defining a conditional probability of $\mathscr{X}$ given $Y^{-1}(\mathscr{Y})$.

Now let $X(i, j) = i$, and define $\mathscr{X}$ and $\mathscr{Y}$ to be the space of functions $f$ that take finitely many non-zero values.

Then $P_Y^{X,Y} W(j) = \sum_{i \geq j} W(i, j)$ where $W$ is any real valued function finitely non-zero.

$$P_Y^{X,Y}(W Y) = P_Y^{X,Y} W(Y) = \sum_{i \geq Y} W(i, Y), \text{ a function in } Y^{-1}(\mathscr{Y}).$$

The quotient probability $P_Y^X f(j) = \sum_{i \geq j} f(i)$ defines a probability on $\mathscr{X}$ for each $j$, such that $\{g(j) P_Y^X f(j)\} \in \mathscr{Y}$ for each $g$ in $\mathscr{Y}$.

$$P_Y^{X,Y} fg = \sum_{i \geq j} f(i) g(j) = g P_Y^X f.$$

Here $X$ and $Y$ are not independent because $P_Y^X f(j)$ varies with $j$.

## 3.4. Marginalization Paradoxes

In Stone and Dawid (1972), and in Dawid, Stone and Zidek (1973) a number of "marginalization paradoxes" are produced using improper priors. See also Sudderth (1980) where it is shown that no marginalization paradox arises with unitary finitely additive priors.

Consider example 1, Stone and Dawid (1972). Random variables $X$ and $Y$ are independent exponential given parameters $\theta\phi$ and $\phi$, and $\theta$, $\phi$ have density $e^{-\theta}$ with respect to lebesgue measure on the positive quadrant. The joint density of $X$, $Y$, $\theta$, $\phi$ is $e^{-\theta}\theta\phi^2 \exp[-\phi(\theta X + Y)]$, and the product rule $P^{X,Y,\theta,\phi} = P^{\theta,\phi}P^{X Y,\theta,\phi}_{\theta,\phi}$ is satisfied.

The conditional density of $\theta$, $\phi$ given $X$, $Y$ is $e^{-\theta}\theta\phi^2 \exp[-\phi(\theta X + Y)]/f(X, Y)$ where $f(X, Y)$ is the density of $X$, $Y$:

$$\iint e^{-\theta}\theta\phi^2 \exp[-\phi(\theta X + Y)]\,d\theta d\phi.$$

Again the product rule is satisfied.

However the conditional density of $\theta$ given $X$, $Z$ where $Z = Y/X$ is $e^{-\theta}\theta/(\theta + Z)^3 f(Z)$ which does not depend on $X$; Stone and Dawid take this to imply that the conditional density of $\theta$ given $Z$ is $e^{-\theta}\theta/(\theta + Z)^3 f(Z)$. Similarly the conditional density of $Z$ given $\theta$ and $\phi$ is $\theta/(\theta + Z)^2$ independent of $\phi$; Stone and Dawid take this to imply the density of $Z$ given $\theta$ is $\theta/(\theta + Z)^2$, which is inconsistent with the conditional density of $\theta$ given $Z$ being $e^{-\theta}\theta/(\theta + Z)^3 f(Z)$.

The paradox is caused by the implicit assumption that $P^{Z,\theta}_{\theta} = P^{Z,\theta,\phi}_{\theta,\phi}$ if $P^{Z,\theta,\phi}_{\theta,\phi}$ is independent of $\phi$; this assumption is valid if $\theta$, $\phi$ given $\theta$ is unitary for then, letting $f$ be a continuous function of two real variables, vanishing outside some square,

$$P^{Z,\theta,\phi}_{\theta}[f] = P^{\theta,\phi}_{\theta}P^{Z,\theta,\phi}_{\theta,\phi}[f]$$
$$= P^{Z,\theta,\phi}_{\theta,\phi}f \text{ (which is independent of } \phi).$$

Thus $P^{Z,\theta}_{\theta}f = P^{Z,\theta,\phi}_{\theta,\phi}f$.

However, if $\theta$, $\phi$ given $\theta$ is not unitary, $P^{\theta,\phi}_{\theta}P^{Z,\theta,\phi}_{\theta,\phi}f$ is not defined, since $P^{\theta,\phi}_{\theta}$ is not defined for non-zero functions constant over $\phi$. Instead,

$$P^{Z,\theta,\phi}_{\theta}[h(\phi)f] = P^{\theta,\phi}_{\theta}h(\phi)P^{Z,\theta,\phi}_{\theta,\phi}f.$$

Thus the joint distribution of $\phi$ and $Z$ given $\theta$ is a product distribution, but because $\phi$ given $\theta$ is not unitary it is not possible to determine the marginal distribution of $Z$ given $\theta$. (In the same way, if $X$, $Y$ is uniformly distributed over the plane we cannot determine that $X$ is uniformly distributed on the line.)

In the example, take $\theta$ to have density $e^{-\theta}$ and $\phi$ given $\theta$ to be uniform. Then $Z$, $\phi$ given $\theta$ has density $\theta/(\theta + Z)^2$; but $Z$ given $\theta$ does not have density $\theta/(\theta + Z)^2$ because it is not valid to integrate over $\phi$.


## 3.5. Bayes Theorem

A real valued function $f$ on $S$ is a density on $\mathcal{X}$ if $fX \in \mathcal{X}$ for $X \in \mathcal{X}$. A probability space $\mathcal{X}$ is $\sigma$-finite if there exists $X_0 \in \mathcal{X}$, $X_0 > 0$; or equivalently, if there exists $X_n \in \mathcal{X}$, $X_n \uparrow 1$ as $n \to \infty$.

**Bayes Theorem.**

*Let $P$ be a probability on $U$, $\mathscr{L}$.*

*Let $X, Y$ and $X \times Y$ be random variables from $U, \mathscr{L}$ to $S, \mathscr{X}, T, \mathscr{Y}$ and $S \times T, \mathscr{X} \times \mathscr{Y}$.*

*Let $f$ be a density on $\mathscr{X} \times \mathscr{Y}$.*

*Let $\mathscr{X} \times \mathscr{Y}$ be $\sigma$-finite.*

*Let $f_T(t):s \to f(s,t) \in \mathscr{X}$ each $t$.*

*Let $f/P^X f_T:(s,t) \to f(s,t)/P[f(X,t)]$ be a density on $\mathscr{X} \times \mathscr{Y}$.*

*Let $P^Y_X g = Q^Y(g f_S)$ for some $Q$ on $\mathscr{Y}$, each $g \in \mathscr{Y}$.*

*$\left[ \text{that is } (P^Y_X g)(s) = Q^{Y^{-1}(\mathscr{Y})}[g(Y)f(s,Y)] \right]$*

*Then*

$P^Y g = Q^Y(g P^X f_T)$ $g \in \mathscr{Y}$.

$P^X_Y h = P^X(h f_T)/P^X[f_T]$ as $P^Y$ each $h \in \mathscr{X}$.

$\left[ \text{that is } (P^X_Y h)(t) = P[h(X)f(X,t)]/Pf(X,t), \text{ except for a set } T_0 \text{ of } t \text{ values with } P^Y T_0 = 0 \right].$

PROOF. Let $h_n \in \mathscr{X}$, $h_n \uparrow 1$, since $\mathscr{X}$ is $\sigma$-finite. For $g \in \mathscr{Y}$, $g \geq 0$, $h_n g \uparrow g$; since $h_n g \in \mathscr{X} \times \mathscr{Y}$ and $g \in \mathscr{Y}$,

$$P^{X,Y} h_n g = P^{(X,Y)^{-1}} \mathscr{X} \times \mathscr{Y}_{h_n}(X)g(Y) \uparrow P^{Y^{-1}(\mathscr{Y})} g(Y) = P^Y g$$

$$P^{X,Y} h_n g = P^X P^Y_X h_n g = P^X[h_n Q^Y(g f_S)]$$

$$P^Y g = P^X Q^Y(g f_S) = Q^Y P^X(g_T f_T) = Q^Y(g P^X f_T).$$

This shows $P^Y g = Q^Y(g P^X f_T)$ for $g \geq 0$, and general $g$ follow easily.

Secondly, it is necessary to show that $P^X_Y$ is a quotient probability; from the given definition,

$$P^{X,Y}_Y[gh] = P^X[g_T h f_T]/P^X[f_T].$$

Then      $P^Y P^{X,Y}_Y gh = P^Y(g P^X h f_T/P^X f_T)$

$\qquad\qquad\qquad = Q^Y(g P^X h f_T)$   from the first part of the proof,

$\qquad\qquad\qquad = P^X Q^Y(g h_S f_S)$

$\qquad\qquad\qquad = P^X[h P^Y_X g]$   from (v)

$\qquad\qquad\qquad = P^{X,Y} gh.$

Thus $P^{X,Y}_Y$ as defined satisfies the product rule, and for any other conditional probability $Q^{X,Y}_Y$ satisfying the product rule,

$$P^Y |P^{X,Y}_Y W - Q^{X,Y}_Y W| = 0.$$

Thus $P^Y g |P^X_Y h - Q^X_Y h| = 0$ for any quotient probability $Q$ satisfying the product rule. Since $\mathscr{Y}$ is $\sigma$-finite, $g$ may be chosen positive, and so $P^Y |P^X_Y h - Q^X_Y h| = 0.$   $\square$

In terms of densities, we have that $Y$ given $X$ has density $f_S$ with respect to some probability $Q^Y$; under specified conditions on $f$ and $\mathscr{X}$, $X$ given $Y$ is a unitary probability with density $f_T/P^X f_T$ with respect to $P^X$. In the usual terminology, $f_S$ would be the *likelihood* of $Y$ given $X$, $P^X$ is the prior distribution of $X$, and $f_T/P^X f_T$ is the *posterior density* of $X$ given $Y$. Frequently $P^X$ has some *prior density* P with respect to a probability $R^X$, and then the conditional probability of $X$ given $Y$ has *posterior density* $f_T p/R^X(f_T p)$ with respect to $R^X$.

Note that $P^X$ and $P_X^Y$ may not be unitary, but under the conditions of the theorem $P_Y^X$ is unitary. Renyi (1970) takes unitary conditional probabilities as the basic concept, expressing non-unitary probabilities such as lebesgue measure by families of such conditional probabilities. It seems simpler to go the other way, to define unitary conditional probabilities from non-unitary probabilities; indeed we allow non-unitary conditional probabilities in general, though our Bayes theorem produces only unitary conditional probabilities.

## 3.6. Binomial Conditional Probability

The binomial distribution is defined for $n$ 0–1 random variables $X_1, X_2, \ldots, X_n$ given a parameter $p$, $0 \leq p \leq 1$, by

$$P[X_i = X_i, i = 1, \ldots, n | p] = p^{\Sigma x_i}(1 - p)^{n - \Sigma x_i}.$$

The random variables $X_1, X_2, \ldots, X_n$ are independent and identically distributed given $p$, with $P(X_i | p) = p$.

In Bayesian analysis $p$ (as well as the $X_i$) is taken to be a random variable on some underlying probability space $\mathscr{X}$ and a probability $P$ is assumed on $\mathscr{X}$ such that the conditional distribution of $X_1, \ldots, X_n$ given $p$ is binomial. If $p$ is not unitary, the marginal probability of $x_1, \ldots, x_n = P(p^{\Sigma x_i}(1 - p)^{n - \Sigma x_i})$ is not defined for all $x$; thus conditional probability given the observations must be carefully handled. For example, if $Pf = \int [f(p)/p(1 - p)]dp$ using Haldane's prior, then $P[X_1 = 0]$ is not defined and the conditional distribution of $P$ given $X_1 = 0$ is not uniquely defined.

Assume that $p^m(1 - p)^{m'} \in \mathscr{X}$ whenever $m \geq a, m' \geq b$. Then define $\mathscr{X}_{a,b}^n$ to be the probability subspace of $\mathscr{X}$ including all functions

$$X_i = x_i, i = 1, \ldots, n', n' > n, x_{n'} = 1, \textstyle\sum x_i = a, n' - \sum x_i \geq b$$
$$X_i = x_i, i = 1, \ldots, n', n' > n, x_{n'} = 0, \textstyle\sum x_i \geq a, n' - \sum x_i = b$$
$$X_i = x_i, i = 1, \ldots, n, \textstyle\sum x_i \geq a, n - \sum x_i \geq b.$$

Thus $\mathscr{X}_{a,b}^n$ corresponds to the shortest sequences of observations containing at least $n$ observations, at least $a$ successes, and at least $b$ failures. The conditional probability of $p$ given $\mathscr{X}_{a,b}^n$ is

$$P[f(p) | \mathscr{X}_{a,b}^n] = \frac{P[f(p)p^{\Sigma X_i}(1 - p)^{n' - \Sigma X_i}]}{P[p^{\Sigma X_i}(1 - p)^{n' - \Sigma X_i}]}.$$

It may be verified that $P[P(f(p) | \mathscr{X}_{a,b}^n)] = Pf(p)$.

Consider, for example, Haldane's prior; here $a = 1$, $b = 1$ since $\int [f(p)/P(1-p)]dp$ exists whenever $f(p) = p^m(1-p)^{m'}$, $m \geq 1$, $m' \geq 1$. (Of course $a$ and $b$ could be positive fractions but this does not change $\mathscr{X}^n_{a,b}$.) Then $\mathscr{X}^2_{1,1}$ is generated from the sequences

$$001, \ 0001, \ 00001, \ \ldots$$
$$110, \ 1110, \ 11110, \ \ldots$$
$$01, \ 10.$$

The conditional probability of $p$ given $\mathscr{X}^2_{1,1}$ is

$$P[f(p)|\mathbf{X}] = \frac{\int f(p)p^{\Sigma X_i - 1}(1-p)^{n' - \Sigma X_i - 1}\,dp}{\int p^{\Sigma X_i - 1}(1-p)^{n' - \Sigma X_i - 1}\,dp}$$

which is defined for each of the specified sequences since $\sum x_i \geq 1$, $n' - \sum x_i \geq 1$. If $a = 0$, $b = 0$ we would have $\mathscr{X}^2_{0,0}$ generated from 00, 01, 10, 11; and the conditional probability of $f(p)$ averages to the probability of $f(p)$ when weighted by the probabilities of 00, 01, 10, 11. In the case $a = 1$, $b = 1$ the sequences 00 and 11 do not have defined probabilities, so the average that validates conditional probabilities is not available—00 is replaced by 001, 0001, ..., and 11 is replaced by 110, 1110, ..., and an average with valid marginal weights becomes available.

In application, we will be able to give conditional probabilities whenever the data sequence has at least one 1 and at least one 0. If the data are of form: all 0's or all 1's, no conditional probability consistent with the axioms is available.

## 3.7. Problems

Problems (Q) are ones that I find very hard.

E1. Let $N$ denote the positive integers, and let $\mathscr{Z}$ be the space of sequences $\{z_1, z_2, \ldots, z_n \ldots\}$ with $\sum |z_i| < \infty$. Let $X$ and $Y$ be random variables into $(N, \mathscr{Z})$, and let $X \times Y$ be a random variable into $N \times N$, with $P\{X = i, Y = j\} = p_{i,j}$. Define $p_j = \sum_i p_{i,j}$. Let $\delta_i(k) = \{i = k\}$, $k \in \mathscr{Z}$.

Then                         $P^X_Y(\delta_i)(j) = \{p_j > 0\}(p_{i,j}/p_j)$.

E2. Let $\mathbb{R}$ be the real line, $\bar{B}$ be the space of bounded lebesgue integrable functions (obtained by completing the probability that values an interval by its length, and accepting bounded functions in that completion). Let $X$ and $Y$ be random variables into $(\mathbb{R}, \bar{B})$ and let $f(x, y)$ be such that $g(y)f(x, y) \in \bar{B} \times \bar{B}$ for each $g \in \bar{B}$.

Suppose         $P^{X,Y}W = \iint W(x, y)f(x, y)dxdy$.
Then                $P^Y g = \iint g(y)f(x, y)dxdy$,
and         $P^X_Y(h)(y) = \int h(x)f(x, y)dx/\int f(x, y)dx$.

P1. Let $X$ be a real valued random variable uniformly distributed, and let the conditional distribution of $Y$ given $X$ give probability $\frac{1}{2}$ to $X - \frac{1}{2}$ and probability $\frac{1}{2}$ to $X + \frac{1}{2}$. Find the joint distribution of $X \times Y$ and the conditional distribution of $X$ given $Y$. (Bayes theorem fails.)

P2. Say $X \sim N(\mu_0, \sigma_0^2)$ if $X$ is a real random variable having density $\exp[-\frac{1}{2}(\mu_0 - x)^2/\sigma_0^2]/(\sigma_0 \sqrt{2\pi})$ with respect to lebesgue measure. Suppose $\theta \sim N(\mu_0, \sigma_0^2)$, $X|\theta \sim N(\theta, \sigma^2)$. Find the distribution of $\theta$ given $X$.
If, in addition $Y|X, \theta \sim N(X - \theta, \sigma^2)$, find $\theta|X, Y$.

P3. The three observations $1, 3, 7$, given $\theta$, are from a normal family $(1/\sqrt{2\pi}) \exp[-\frac{1}{2}(x - \theta)^2]$ with probability $\frac{1}{4}$ or from the family $\frac{1}{2} \exp[-|x - \theta|]$ with probability $\frac{3}{4}$. The prior distribution of $\theta$ is uniform. Find the posterior distribution of $\theta$ given the observations, and the posterior probability that the normal is the true distribution.

P4. Assume that $X|\theta \sim N(\theta, 1)$, and that $\theta \sim N(\theta_0, \sigma_0^2)$. Usually $\theta_0$ and $\sigma_0^2$ are assumed known, but suppose, as an afterthought, you decide that $\theta_0 \sim N(0, \sigma_1^2)$. Find the posterior distribution of $\theta$ given $X$.

P5. Let $X_1, \ldots, X_n$ be a sample from the uniform $(\theta - \frac{1}{2}, \theta + \frac{1}{2})$ given $\theta$. Let the prior probability of $\theta$ be uniform. Find the posterior probability of $\theta|X_1, \ldots, X_n$ and compute the posterior mean and variance.

Q1. Suppose $\mathscr{X}$ and $\mathscr{Y}$ are probability spaces on $S$, and that $P: \mathscr{X}$ to $\mathscr{Y}$ satisfies the axioms of conditional probability, but that $\mathscr{X} \supset \mathscr{Y}$ is not assumed. When is it possible to extend $P$ to $\mathscr{Z}$ including $\mathscr{X}$ and $\mathscr{Y}$, so that $P: \mathscr{Z}$ to $\mathscr{Y}$ is a conditional probability?

Q2. If $P$ is a conditional probability on $\mathscr{X}$ to $\mathscr{Y}$, does there exist a complete conditional probability $P'$ on $\mathscr{X}'$ to $\mathscr{Y}$ such that $P'$ coincides with $P$ on $\mathscr{X} \subset \mathscr{X}'$?

## 3.8. References

Dawid, A. P., Stone, M. and Zidek, J. V. (1973), Marginalization paradoxes in Bayesian and statistical inference, *J. Roy. Stat. Soc.* B **35**, 189–223.

Kolmogorov, A. N. (1933), *Foundations of the Theory of Probability*. New York: Chelsea.

Renyi, A. (1970), *Probability Theory*. New York: American Elsevier.

Stone, M. and Dawid, A. P. (1972), Un-Bayesian implications of improper Bayes inference in routine statistical problems, *Biometrika* **59**, 369–373.

Sudderth, W. D. (1980), Finitely additive priors, coherence, and the marginalization paradox, *J. Roy. Stat. Soc.* B **42**, 339–341.

Tjur, T. (1972), *On the mathematical foundations of probability*. Inst. of Math. Statist., University of Copenhagen.

———(1974), *Conditional probability distribution*. Inst. of Math. Statist., University of Copenhagen.

# CHAPTER 4
# Convergence

## 4.0. Introduction

Notions of convergence, as the amount of information increases, are necessary to check the consequences of probability assignments in empirical fact. For example, if we assume that a penny has probability 1/2 of coming down heads on each toss, and that the different tosses are independent, it follows that the limiting proportion of heads will be 1/2 almost surely.

A standard method of evaluating statistical procedures is through their asymptotic properties; partly this is a matter of necessity because the asymptotic behavior is simple. For example, one of the desirable properties of a maximum likelihood estimate is that it converges, in a certain sense under certain regularity conditions, to the unknown parameter value. A famous theorem due to Doob (1949) handles consistency of Bayes procedures—if any estimate converges to the unknown parameter value in probability then the posterior distribution concentrates on the unknown value as the data increases.

## 4.1. Convergence Definitions

Let $\mathscr{X}$ be a probability space on $S$. A real valued function $X$ on $S$ is *measurable* if $\{X > a\}, \{X < -a\} \in \mathscr{X}$ for each $a > 0$. Thus $X$ is a random variable into $R, \mathscr{B}_0$ where $\mathscr{B}_0$ is the smallest probability space containing intervals $(a, \infty)$, $(-\infty, -a)$ for each $a > 0$.

The space of measurable functions is a probability space which includes $\mathscr{X}$. In the following $X, X_1, \ldots, X_n$, are assumed to be measurable functions

34

with respect to $\mathcal{X}$ on $S$, and it is assumed that there is a probability $P$ on $\mathcal{X}$ on $S$.

$X_1 = X_2$ as $P$ means $P(X_1 \neq X_2) = 0$.
$X_n \to X$ as $P$ means $P\{s \mid X_n(s) \not\to X(s)\} = 0$; equivalently, for each $\varepsilon > 0$,
$P\{\mid X_n - X \mid > \varepsilon \text{ some } n > N\} \to 0$ as $N \to \infty$.
$X_n \to X$ in $P$ means $P\{\mid X_n - X \mid > \varepsilon\} \to 0$ each $\varepsilon > 0$.
$X_n \to X$ by $P$ means $P \mid X_n - X \mid \to 0$ as $n \to \infty$.

If $X_n \to X$ by $P$ or $X_n \to X$ as $P$ then $X_n \to X$ in $P$.
If $X \in \mathcal{X}$, $X_n \in \mathcal{X}$, $X_n \to X$ in $P$, and $\sup_n P(\mid X_n \mid (\{\mid X_n \mid > A\} + \{\mid X_n \mid < 1/A\})) \to 0$ as $A \to \infty$, then $X_n \to X$ by $P$. The sup condition is a generalization of the notion of uniform integrability, necessary to handle non-unitary $P$. With some mess, it may be shown that it is sufficient to prove the result for $X = 0$;

$$\overline{\lim_n} \, P \mid X_n \mid \leq \sup_n P\left( \mid X_n \mid \left( \{\mid X_n \mid > A\} + \left\{\mid X_n \mid > \frac{1}{A}\right\}\right)\right)$$
$$+ \, \overline{\lim} \, P\left[ \mid X_n \mid \left(\frac{1}{A} \leq \mid X_n \mid < A \right)\right]$$

Since $\overline{\lim} \, P(\mid X_n \mid \{1/A \leq \mid X_n \mid < A\}) \leq A \, \overline{\lim} \, P\{\mid X_n \mid \geq 1/A\} = 0$ all $A$, $\overline{\lim} \, P \mid X_n \mid = 0$.

$X_n \to X$ in $D$ (in distribution) if $P[f(X_n)] \to P[f(X)]$ for each bounded continuous $f$ that vanishes near zero. It is not necessary that the $X_n$ and $X$ be defined on the same probability space; the definition involves only the distribution of the $X_n$ and $X$.

If $X_n \to X$ in $P$ then $X_n \to X$ in $D$. To show this, for each $f$ there is a fixed $\varepsilon_0$ such that $f(x) = 0$ for $\mid x \mid \leq \varepsilon_0$, and for each $k$, $\delta$ there is an $\varepsilon < \varepsilon_0/2$ depending on $(k, \delta)$ such that $\mid f(x) - f(y) \mid < \delta$ for $\mid x \mid < k$, $\mid x - y \mid < \varepsilon$. Thus

$$\mid f(x) - f(y) \mid \leq 2 \sup f(\{\mid x \mid > k\} + \{\mid x - y \mid > \varepsilon\}) + \delta\{\mid x \mid \geq \varepsilon_0/2\}$$
$$\overline{\lim} \mid f(X) - f(X_n) \mid \leq 2 \sup f P\{\mid X \mid > k\} + \delta P(\mid X \mid \geq \varepsilon_0/2)$$

Choosing $k$ large and $\delta$ small gives $P \mid f(X) - f(X_n) \mid \to 0 \Rightarrow X_n \to X$ in $D$.

## 4.2. Mean Convergence of Conditional Probabilities

**Theorem.** *Let $\mathcal{X}_n$ be a sequence of probability spaces contained in the probability space $\mathcal{X}$, let $P$ be a probability on $\mathcal{X}$, and let $P_n = P^{\mathcal{X}}_{\mathcal{X}_n}$ be the conditional probability of $\mathcal{X}$ given $\mathcal{X}_n$. Let $X \in \mathcal{X}$.*
*(i) If $X$ is mean-approximable by $\mathcal{X}_n$ (that is $P \mid X_n - X \mid \to 0$ for $X_n \in \mathcal{X}_n$), then $X$ is mean-approximable by the sequence $P_n X$.*
*(ii) If $X$ is square-approximable by $\mathcal{X}_n$ (that is $P(X_n - X)^2 \to 0$ for $X_n \in \mathcal{X}_n$), then $X$ is square-approximable by the sequence $P_n X$.*

PROOF. Since $\mathcal{X}_n \subset \mathcal{X}$, $P_n X_n = X_n$ for each $X_n$ in $\mathcal{X}_n$, so $P_n P_n = P_n$.

$$P_n|X - P_n X| \leq P_n|X - X_n| + P_n|X_n - P_n X| = P_n|X - X_n| + P_n|P_n(X_n - X)|$$
$$\leq P_n|X - X_n| + P_n P_n|X_n - X| = 2P_n|X - X_n|.$$

Thus if $X$ is mean-approximable by $\mathcal{X}_n$, it is mean-approximable by $P_n X$.

$$P_n(X - X_n)^2 = P_n(X - P_n X)^2 + 2P_n[(X - P_n X)(P_n X - X_n)] + P_n(X_n - P_n X)^2$$
$$= P_n(X - P_n X)^2 + 0 + P_n(X_n - P_n X)^2.$$

Thus if $X$ is square-approximable by $\mathcal{X}_n$ it is square-approximable by $P_n X$. $\square$

EXAMPLE.

Let $P$ be a complete probability on $\mathcal{G}$ on $U$.
Let $\theta, X, X_1, \ldots, X_n$ be random variables on $U, \mathcal{G}$.
Suppose $X_1, \ldots, X_n, \ldots$ are independent and identically distributed given $\theta$; let $P_\theta^{X_i} = P_\theta^X$ for $i = 1, 2, \ldots$.
Assume that $X$ is a real valued random variable such that $X, X^2 \in \mathcal{G}$. Set $\mu = P(X|\theta)$.

Since $X_1, \ldots, X_n$ are independent given $\theta$,

$$P\left[\left(\frac{1}{n}\sum X_i - \mu\right)^2 \Big| \theta\right] = P[(X - \mu)^2|\theta]/n$$

$$P\left[\frac{1}{n}\sum X_i - \mu\right]^2 = P[X - \mu]^2/n.$$

Note that $\mu^2 \in Z$ because $\mu^2\{|\mu| \leq k\} \uparrow \mu^2$ as $k \uparrow \infty$, $\mu^2\{|\mu| \leq k\} = |\mu|(|\mu| \wedge k) \in \mathcal{G}$, and $P\mu^2\{|\mu| \leq k\} \leq PX^2$ all $k$.

Thus $\mu$ is square-approximable by $1/n \sum X_i$ and hence by $P[\mu|X_1, X_2, \ldots, X_n]$. This is a Bayesian adaptation of the law of large numbers, in which the unknown population mean $\mu$ is increasingly well approximated by its best estimate from the sample $X_1, \ldots, X_n$.

## 4.3. Almost Sure Convergence of Conditional Probabilities

**Theorem.** *Let $P$ be a probability on $\mathcal{X}$, and let $\mathcal{X}_n$ be an increasing sequence of subspaces of $\mathcal{X}$. Let $P_n = P_{\mathcal{X}_n}^{\mathcal{X}}$. Let $X \in \mathcal{X}$. If $X$ is mean or square-approximable by $\{\mathcal{X}_n\}$, then*

$$P_n X \to X \text{ as } P.$$

PROOF. Let $X_n = P_n X$. Then $P_{n-1} X_n = X_{n-1}$ by the product law, so $X_n$ is a martingale when $P$ is unitary; in this case the theorem is well known, Doob (1953).

**Lemma.**
$$P\{\sup_{1 \leq i \leq n} |X_i| \geq \varepsilon\} \leq PX_n^2/\varepsilon^2 \text{ if } X_n^2 \in \mathcal{X},$$
$$P\{\sup_{1 \leq i \leq n} |X_i| \geq \varepsilon\} \leq P|X_n|/\varepsilon.$$

PROOF. Let $A_i = \{|X_1| < \varepsilon, |X_2| < \varepsilon, \ldots, |X_i| \geq \varepsilon\}, i = 1, 2, \ldots, n.$
Then

$$\sum_{i=1}^{n} A_i = \{\sup_{1 \leq i \leq n} |X_i| \geq \varepsilon\}.$$
$$A_i X_n^2 = A_i(X_i^2 + 2(X_n - X_i)X_i + (X_n - X_i)^2)$$
$$P[A_i(X_n - X_i)X_i] = PP_i[A_i X_i(X_n - X_i)] = P[A_i X_i P_i(X_n - X_i)] = 0$$
$$P(A_i X_n^2) \geq \varepsilon^2 PA_i \text{ since } |X_i| \geq \varepsilon \text{ when } A_i \neq 0$$
$$PX_n^2 \geq \varepsilon^2 P \sum A_i = \varepsilon^2 P\{\sup_{1 \leq i \leq n} |X_i| \geq \varepsilon\} \text{ as required.}$$

For the second result, define $B_i = \{X_1 < \varepsilon, X_2 < \varepsilon, \ldots, X_i \geq \varepsilon\}$

Then
$$\sum B_i = \{\sup_{1 \leq i \leq n} X_i \geq \varepsilon\}$$
$$X_n B_i = B_i(X_i + X_n - X_i)$$
$$PX_n B_i = PB_i X_i + PP_i B_i(X_n - X_i) = PB_i X_i$$
$$\geq \varepsilon PB_i$$
$$P(X_n^+) \geq P(X_n \sum B_i) \geq \varepsilon P(\sum B_i) = \varepsilon P\{\sup X_i \geq \varepsilon\}$$
$$- P(X_n^-) \geq \varepsilon P\{\inf X_i \leq -\varepsilon\}$$
$$P|X_n| \geq \varepsilon P\{\sup |X_i| \geq \varepsilon\}.$$

Turning to the theorem, if $X$ is square-approximable, from 4.2, $P(X_n - X)^2 \to 0$, and for a suitable subsequence $\sum P(X_{n_r} - X)^2 < \infty$.

$$\{\sup_{i, j \geq N} |X_i - X_j| \geq 4\varepsilon\} \leq \sup(\{\sup_{\substack{n_r \geq N \\ n_{r-1} \leq i \leq n_r}} |X_i - X_{n_r}| \geq \varepsilon\} + \{|X - X_{n_r}| \geq \varepsilon\})$$
$$P\{\sup_{i, j \geq N} |X_i - X_j| \geq 4\varepsilon\} \leq \sum_{n_r \geq N} [P(X_{n_r} - X_{n_{r-1}})^2/\varepsilon^2 + P(X - X_{n_r})^2/\varepsilon^2]$$
$$\to 0 \text{ as } N \to \infty,$$

using the lemma and $\sum P(X_{n_r} - X)^2 < \infty$.
If $X$ is mean approximable, the same argument holds with $P|X_{n_r} - X|$ replacing $P(X_{n_r} - X)^2$ and $P|X_{n_r} - X_{n_{r-1}}|$ replacing $P(X_{n_r} - X_{n_{r-1}})^2$. □

## 4.4. Consistency of Posterior Distributions

**Doob's Theorem.** *Let $\mathscr{X}_n$ be an increasing sequence of subprobability spaces of $\mathscr{X}$.*

*Let $P$ be a probability on $\mathscr{X}$, let $P_n$ be conditional probabilities of $\mathscr{X}$ given $\mathscr{X}_n$.*

*Let $\theta$ be a real valued random variable such that $\{a \leqq \theta \leqq b\} \in \mathscr{X}$ for $a, b$ finite, and suppose that $\theta$ is approximable by $\mathscr{X}_n : X_n \to \theta$ in $P$ some $X_n$ in $\mathscr{X}_n$.*

*Let $C_n(a, b) = P_n\{a \leqq \theta \leqq b\}$.*

*Then $C_n(\theta - \varepsilon, \theta + \varepsilon) \to 1$ as $P$.*

PROOF. Doob (1949) proved a version of this theorem for $P$ unitary. See also Schwartz (1965) and Berk (1970).

Let $a$ and $b$ not be atoms of $\theta : P\{a = \theta\} = P\{b = \theta\} = 0$. $|\{a \leqq \theta \leqq b\} - \{a \leqq X_n \leqq b\}| \leqq \{|a - \theta| \leqq \varepsilon\} + \{|b - \theta| \leqq \varepsilon\} + \{|X_n - \theta| \geqq \varepsilon\}$. Since $P\{|X_n - \theta| \geqq \varepsilon\} \to 0$ as $n \to \infty$, and $P\{|a - \theta| \leqq \varepsilon\} + P\{|b - \theta| \leqq \varepsilon\} \to 0$ as $\varepsilon \to 0$, $P\{a \leqq \theta \leqq b\} - \{a \leqq X_n \leqq b\}| \to 0$.

From 4.3, $P_n\{a \leqq \theta \leqq b\} \to \{a \leqq \theta \leqq b\}$ as $P$.

$$C_n(a, b)(s) \to \{a \leqq \theta(s) \leqq b\} \text{ except for } s \in A, PA = 0.$$

$$C_n(\theta(s) - \varepsilon, \theta(s) + \varepsilon) \to 1 \text{ except for } s \in A.$$

$$C_n(\theta - \varepsilon, \theta + \varepsilon) \to 1 \text{ as } P. \qquad \square$$

The condition of approximability and the conclusion of consistency may both be expressed conditionally on $\theta$; $\theta$ is approximable by $\mathscr{X}_n$ if $P_\theta[|X_n - \theta| > \varepsilon] \to 0$ except for a set of $\theta$-values of probability zero; and $C_n(\theta - \varepsilon, \theta + \varepsilon) \to 1$ as $P$ implies that $C_n(\theta - \varepsilon, \theta + \varepsilon) \to 1$ as $P_\theta$, except for a set of $\theta$ values of probability zero.

EXAMPLE. Let $X_1, X_2, \ldots, X_n$ be a random sample from $N(\theta, 1)$; let $\theta$ be uniformly distributed on the line. Let $\mathscr{X}_n$ denote the probability space generated by $X_1, \ldots, X_n$ and note that $|\bar{X}_n - \theta| \to 0$ in probability given $\theta$. The conditional distribution of $\theta$ given $\mathscr{X}_n$ is unitary. Therefore it concentrates on the true value $\theta$ as $n \to \infty$, except for a set of $\theta$ values of probability zero.

## 4.5. Binomial Case

Let $P$ be a probability on $\mathscr{X}$ on $S$.

Let $P$, the binomial parameter, be a real valued random variable, $0 \leqq p \leqq 1$, defined on $\mathscr{X}$.

Let $X_1, X_2, \ldots, X_n$ be $n$ Bernoulli random variables, each taking the values 0 or 1; thus $X_1, X_2, \ldots, X_n$ maps $S$ to the space of *n-tuples* $(x_1, x_2, \ldots, x_n)$ where $x_i = 0$ or 1.

A binomial distribution has $X_1, X_2, \ldots, X_n$ independent and identically distributed given $p$:

$$P[X_i = x_i, i = 1, \ldots, n | p] = p^{\Sigma x_i}(1 - p)^{n - \Sigma x_i}.$$

The posterior distribution of $p$ given $X_1, \ldots, X_n$ is

$$P_n f = P[f(p) | X_1, \ldots, X_n] = P[f(p)p^{\Sigma X_i}(1 - p)^{n - \Sigma X_i}] / P(p^{\Sigma X_i}(1 - p)^{n - \Sigma X_i})$$

defined whenever $p^{\Sigma X_i}(1 - p)^{n - \Sigma X_i} \in \mathcal{X}$.

(Note that the conditioning space $\mathcal{X}_n$ must contain observations of varying length if $P$ is not unitary.)

**Theorem.** *Let $p, X_1, \ldots, X_n$ be random variables on $\mathcal{X}$, and let $X_1, \ldots, X_n, \ldots$ be Bernoulli given $p$. Assume that $p^m(1 - p)^{m'} \in \mathcal{X}$ for $m \geq a, m' \geq b$. Say that $p_0$ is in the support of $P$ if $P[|p - p_0| < \varepsilon] > 0$ for each $\varepsilon > 0$. Then $P_n[|p - p_0| < \varepsilon] \to 1$ as $P_{p_0}$ if and only if $p_0$ is in the support of $P$.*

PROOF. Note that $P_n[|p - p_0| < \varepsilon]$ is a function in $\mathcal{X}_n$. Let $R = \Sigma X_i$. If $p_0$ is not in the support of $P$, $P[|p - p_0| < \varepsilon] = 0$ some $\varepsilon > 0$, and so $P_n[|p - p_0| < \varepsilon] = 0$ for all $n$. This establishes the "only if."

Now suppose $p_0$ lies in the carrier of $P$. Assume $0 < p_0 < 1$.

Let $f(p_0, p) = p_0 \log p + (1 - p_0) \log(1 - p)$.

Then $f(p_0, p)$, as a function of $p$, is continuous and has a unique maximum at $p = p_0$. Thus for each small $\delta > 0$, there exists $\Delta > 0$ with

$$f(p_0, p_1) > f(p_0, p_2) + \Delta \quad \text{whenever} \quad |p_0 - p_1| < \delta,$$
$$|p_0 - p_2| > 2\delta.$$

As $n \to \infty$, $R/n \to p_0$ as $P_{p_0}$ from the strong law of large numbers.

$$f\left(\frac{R}{n}, p_1\right) > f\left(\frac{R}{n}, p_2\right) + \Delta \quad \text{whenever} \quad |p_0 - p_1| < \delta,$$
$$|p_0 - p_2| > 2\delta,$$

for all large $n$, with probability approaching 1 as $n \to \infty$. (Note that the conditioning space may include observation sequences of length greater than $n$, but the inequality holds for all these sequences.)

$$P_n\{|p - p_0| \leq \delta\} / P_n\{|p - p_0| \geq 2\delta\}$$
$$= P\left[ \{|p - p_0| \leq \delta\} \exp\left[ nf\left(\frac{R}{n}, p\right)\right]\right] \Big/$$
$$P\left[ \{|p - p_0| \geq 2\delta\} \exp\left[ nf\left(\frac{R}{n}, p\right)\right]\right]$$
$$\geq e^{n\Delta} P\{|p - p_0| < \delta\} / P\{|p - p_0| > 2\delta\} \to \infty \quad \text{as} \quad P_{p_0}.$$

Thus $P_n\{|p - p_0| \geq 2\delta\} \to 0$ as $P_{p_0}$ as required.                    □

*Remark*: The same result generalizes to multinominal distributions, but not to observations carried by a countable number of points, as shown by Freedman (1963).

## 4.6. Exchangeable Sequences

Let $P$ be a probability on a space $\mathscr{X}$. A sequence of random variables $X_1$, $X_2, \ldots, X_n, \ldots$ defined on $\mathscr{X}$ is said to be *exchangeable* if $X_1, X_2, \ldots, X_n$ has the same distribution as $X_{\sigma 1}, X_{\sigma 2}, \ldots, X_{\sigma n}$ for each $n$ and each permutation $\sigma$.

**De Finetti's Theorem.** *Let* $X_1, X_2, \ldots, X_n, \ldots$ *be an exchangeable sequence of* $0 - 1$ *random variables on a probability space* $\mathscr{X}$ *with unitary probability* $P$. *Then* $X_1, X_2, \ldots, X_n, \ldots$ *are conditionally independent and identically distributed given the random variable* $p = \lim \sum_{i=1}^{2^n} X_i/2^n$ *if the limit exists,* $p = 0$ *if the limit fails.*

PROOF. Let $\bar{X}_n = (1/n)\sum_{i=1}^{n} X_i$.

Then $P(\bar{X}_n - \bar{X}_m)^2 = [P(X_1^2) - P(X_1 X_2)](m - n)/mn$

$$\sum P|\bar{X}_{2^{s+1}} - \bar{X}_{2^r}| \leq \infty$$

$$P\{|\bar{X}_{2^{r+1}} - \bar{X}_{2^N}| > \varepsilon\} \leq P\left\{\sum_{r=N}^{M} |\bar{X}_{2^{r+1}} - \bar{X}_{2^r}| > \varepsilon\right\} \to 0 \text{ as } N, M \to \infty.$$

Thus $p = \lim \sum_{i=1}^{2^n} X_i/2^n$ is defined except for a set $A$ of probability 0; define $p = 0$ on the set $A$. Define a conditional probability on $X_1, X_2, \ldots, X_n, \ldots$ given $p$ by $P[1|p] = 1, P[X_i|p] = p$, and let the $X_1, X_2, \ldots, X_n, \ldots$ be independent and identically distributed given $p$. The specified probability obviously satisfies the axioms of conditional probability except for possibly axiom 1 and the product axiom.

For axiom 1 we need show that for Baire functions $h$, and for functions $g$ in the smallest probability space including $X_1, \ldots, X_n$, $P[h(p)g|p] = h(p)P[g|p]$.

Take $h$ to be continuous and bounded, and note that the class of functions $g$ satisfying the identity is a limit space, so that we need only verify that $P[h(p)\prod X_i|p] = h(p)P[\prod X_i|p]$. (The probability space generated by $X_1, X_2, \ldots, X_n$ is the smallest limit space including $X_{i_1}, X_{i_2}, \ldots, X_{i_n}$ for each sequence $i_1, i_2, \ldots, i_n$. If the axiom is satisfied for $X_1, X_2, \ldots, X_n$, by symmetry it is satisfied for $X_{i_1}, X_{i_2}, \ldots, X_{i_n}$.) Define $\bar{X}_{2^r,i} = \sum X_j \times \{2^{r(i-1)} < j \leq 2^{ri}\}/[2^{ri} - 2^{r(i-1)}]$. Then $\bar{X}_{2^r,i} \to p$ as $r \to \infty$, whenever $\bar{X}_{2^r} \to p$ as $r \to \infty$. If $\bar{X}_{2^r} \to p$,

$$P[h(p)\prod X_i|p] = P[\lim h(\bar{X}_{2^r})\prod X_i|p]$$
$$= P[\lim h(\bar{X}_{2^r})\prod \bar{X}_{2^r,i}|p]$$
$$= h(p)p^n \quad \text{by the law of large numbers}$$
$$= h(p)P[\prod X_i|p].$$

If the limit $\bar{X}_{2^r}$ does not exist, then $p = 0$ and $P[h(p)\prod X_i|p] = 0$. The axiom is true for bounded and continuous $h$, and it is true on a limit space of functions, so it is true for Baire functions.

For the product axiom, proceeding as before requires that

$$PP(\prod X_i | p) = P(\prod X_i).$$

Now
$$P(\prod X_i) = P(\prod \bar{X}_{2^r,i}) = P(\lim_r \prod \bar{X}_{2^r,i}) = P(p^n)$$
$$= P[P(\prod X_i | p)].$$

Thus the product axiom holds concluding the proof. □

*Notes:* This theorem, proved first by de Finetti, has philosophical importance in relating Bayes and frequentist theory. In the frequentist approach, probabilities are based on a sequence of observations $X_1, X_2, \ldots, X_n, \ldots$ that have limiting frequency $p$. De Finetti argues that this form of evidence requires the judgement that the $X_1, X_2, \ldots, X_n, \ldots$ are exchangeable; then the limiting frequency $p$ exists except for a set of probability zero. Thus the frequentist probability statements may be derived from Bayes theory.

The generalization is straightforward for real valued random variables; in this case the conditioning random variables are the functions $\lim \sum_{i=1}^{2^r} \{X_i \leq x\}/2^r$ for rational $x$. The "$\prod X_i$" for generating the probability space including $X_1, X_2, X_3, \ldots$ are the functions $\prod_{i=1}^{n} \{X_i \leq x_i\}$ where the $x_i$ are rational. See for example, Loeve (1955), p. 365.

How can the theorem be generalized to non-unitary probabilities? Consider a special case, corresponding to the prior density proportional to $1/p(1-p)$.

Let $S$ be the set of all $0-1$ sequences. Consider the smallest probability space $\mathscr{X}$ on $S$ that includes all functions $X$ depending on a finite number of elements of the sequence $s$, such that $X(s) = 0$ for $s = 0, 0, 0, \ldots$ or $s = 1, 1, 1, \ldots$. For example the finite sequences containing at least one 1 and one 0 will lie in this space.

Let the sequence $x_1, x_2, \ldots, x_n$ have probability $1 \Big/ \binom{n-1}{\sum x_i - 1}$ for $0 < \sum x_i < n$. This assignment of probability satisfies the axioms. For example $1 = P(01) = P(011) + P(010)$.

The function $\bar{x}_n = (1/n)\sum x_i$ does not lie in the probability space, but the functions $\bar{x}_n - \bar{x}_m$ do, since they give value 0 to $s \equiv 0$ or $s \equiv 1$. Since $P(\bar{x}_n - \bar{x}_m)^2 = \frac{1}{2}P(x_1 - x_2)^2 |n - m|/mn$, the sequence $\bar{x}_{2^n}$ converges except on a set of probability zero to a function, $p$ say. Let $\mathscr{Y}$ be the probability space generated by the polynomials $p^{k_1}(1-p)^{k_2}$ where $k_1, k_2 > 0$. Define a conditional probability on $\mathscr{X}$ given $\mathscr{Y}$ by $P[x_1, x_2, \ldots, x_n | \mathscr{Y}] = p^{\sum x_i}(1-p)^{n - \sum x_i}$.

This conditional probability satisfies the product axiom:

$$PP[x_1, x_2, \ldots, x_n | \mathscr{Y}] = P[p^{\sum x_i}(1-p)^{n-\sum x_i}]$$
$$= P[\lim \bar{X}_{2^N}^{\sum x_i}(1 - \bar{X}_{2^N})^{n - \sum x_i}]$$
$$= P(x_1, x_2, \ldots, x_n).$$

And different sequences $x_1, x_2, \ldots, x_n$ and $x_{n+1}, x_{n+2}, \ldots, x_N$ are independent given $p$.

## 4.7. Problems

E1. In the binomial case the prior $P$ is carried by the rationals $(0, 1]$, with $P(r) = \log[1 - 2^{-N}]/[-N]$, where $N$ is the first integer for which $rN$ is an integer. If $p_0$ is the true value of $p$, show that the posterior distribution of $p$ concentrates on $p_0$ as $n \to \infty$.

E2. Two parameters $U, V$ are independent uniforms. A coin is tossed giving heads with probability $|U - V|$. Find the posterior distribution of $U, V$ given $r$ heads in $n$ tosses, and specify its behavior as $n \to \infty$.

P1. In the binomial case, assume $P\{\tfrac{1}{4}\} = P\{\tfrac{3}{4}\} = \tfrac{1}{2}$. Find the asymptotic behavior of $P_n$, as the true probability $p_0$ ranges over $(0, 1)$. Generalize results to $P$ carried by a finite number of points.

P2. Let $p$ be a binomial parameter, and let $\theta = \tfrac{1}{2}p\{0 \le p \le \tfrac{1}{2}\} + (\tfrac{1}{2} + \tfrac{1}{2}p)\{p > \tfrac{1}{2}\}$. Let $P$ be the prior distribution on $\theta$ which is uniform over $(0, \tfrac{1}{4}) \cup (\tfrac{3}{4}, 1)$. Let $P_n$ be the posterior distribution of $\theta$ after $R$ successes in $n$ trials, and consider its distribution as $R$ varies given the true value $\theta_0 = \tfrac{1}{4}$. Show that $P_n$'s distribution converges to a distribution over the two point distributions carried by $\{\theta = \tfrac{1}{4}\}$ and $\{\theta = \tfrac{3}{4}\}$, with $P\{\theta = \tfrac{1}{4}\}$ uniformly distributed between 0 and 1.

P3. In the binomial case, show that for a particular $\varepsilon > 0$, it may happen that $P(p_0 - \varepsilon, p_0 + \varepsilon) > 0$ but $P_n(p_0 - \varepsilon, p_0 + \varepsilon) \to 0$ as $P_{p_0}$.

P4. In the binomial, let the probability $P$ be defined by

$$P(f) = \int_0^1 f/[p(1 - p)]dp.$$

Show that for each $p_0, 0 \le p_0 \le 1, P_n(p_0 - \varepsilon, p_0 + \varepsilon) \to 1$ as $P_{p_0}$.

P5. The observation $X$ is poisson with parameter $\lambda$, where $P\{\lambda = 1 + 1/i\} = 2^{-i}$, $1 \le i < \infty$. Given $\lambda = 1$, specify the asymptotic behavior of the posterior distribution of $\lambda$.

P6. Let $X$ be $0 - 1$ with probability given the parameter $p, 0 \le p \le 1$,

$$P[X|p] = p + \tfrac{1}{2}\{p = \tfrac{1}{4}\} - \tfrac{1}{2}\{p = \tfrac{3}{4}\}.$$

Let the prior distribution of $p$ be uniform. Show that the posterior distribution given $n$ i.i.d. observations $X_1, X_2, \ldots, X_n$ is not consistent for $p = \tfrac{1}{4}$ or $p = \tfrac{3}{4}$.

P7. Let $P$ be a unitary probability on $\mathcal{X}$. Let $Y$ be a random variable, let $\mathcal{Y}$ be the probability space containing sets of form $\{Y \le a\}$, and let $\mathcal{Y}_n$ be the probability space generated by sets of form $\{Y \le k/2^n\}$, $-\infty < k < \infty$.

Let

$$S_{nk} = \left\{ \frac{k}{2^n} < Y \le \frac{k+1}{2^n} \right\}. \quad \text{Show that}$$

$$P_n(X) = \sum_{P(S_{nk}) > 0} S_{nk} P(S_{nk}X)/P(S_{nk})$$

is a conditional probability on $\mathscr{X}$ to $\mathscr{Y}_n$. Suppose that $P_\infty$ is a conditional probability on $\mathscr{X}$ to $\mathscr{Y}$ such that $P_\infty P_n = P_\infty$. Then $P_n(X) \to P_\infty(X)$ in $P$. (In this way, the conditional probability of $\mathscr{X}$ given the random variable $Y$ is approximated by directly computed conditional probabilities given discrete random variables $Y_n$ which approximate $Y$.)

P8. A Markov chain, with a finite number of states, has probability $p_i$ for the initial state to be state $i$, and probability $p_{ij}$ for a transition from state $i$ to state $j$. The initial state and an infinite sequence of transitions is observed. Assume that the prior distribution for $\{p_i\}$ and $\{p_{ij}\}$ is uniform over the sets $0 \leqq p_i \leqq 1$, $\sum p_i = 1$; $0 \leqq p_{ij} \leqq 1$, $\sum_j p_{ij} = 1$. Specify the limiting posterior distribution of $\{p_i\}$ and $\{p_{ij}\}$.

P9. Let $P$ be a probability on $\mathscr{X}$ such that $X \in \mathscr{X} \Rightarrow X^2 \in \mathscr{X}$. Let $\mathscr{Y}$ be a probability subspace of $\mathscr{X}$ such that $PY^2 = 0 \Rightarrow Y = 0$, and suppose that $\mathscr{Y}$ is complete with respect to the metric $\rho(Y_1, Y_2) = P(Y_1 - Y_2)^2$: if $\rho(Y_n, Y_m) \to 0$, there exists $Y \in \mathscr{Y}$ such that $\rho(Y, Y_n) \to 0$. Define $P(X \mid \mathscr{Y}) = Y$ if $Y \in \mathscr{Y}$ minimizes $P(X - Y)^2$. Show that $P(X \mid \mathscr{Y})$ is uniquely defined, and is a conditional probability on $\mathscr{X}$ to $\mathscr{Y}$. (Doob, 1953).

P10. In the binomial case, if $P$ has an atom at $p_0$, then $P_n\{p_0\} \to 1$ as $P$ if $p_0$ is true.

## 4.8. References

Berk, Robert H. (1970), Consistency a posteriori, *Ann. of Math. Statist.* **41**, 894–906.

Doob, J. L. (1949), Application of the theory of martingales, *Colloques Internationaux du Centre National de la Recherche Scientifique*. Paris, 23–27.

Doob, J. L. (1953), *Stochastic Processes*. New York: John Wiley.

Freedman, David (1963), On the asymptotic behaviour of Bayes estimates in the discrete case, *Ann. of Math. Statist.* **34**, 1386–1403.

Loève, Michel (1955), *Probability Theory*. Princeton: Van Nostrand.

Schwartz, L. (1965), On Bayes procedures, *Z. Wahr.* **4**, 10–26.

# CHAPTER 5
# Making Probabilities

## 5.0. Introduction

The essence of Bayes theory is giving probability values to bets. Methods of generating such probabilities are what separate the various theories.

If probabilities are personal opinions, then they are determined by asking questions or observing which of a family of bets an individual accepts. There is a small theory for extracting personal probabilities, the elicitation of probabilities, for example Winkler (1967). To discover a person's probabilities for the disjoint events $A_i$ for example, you offer to pay $+ \log x_i$ if $A_i$ occurs where $x_i$ is the person's stated probability for the event $A_i$. If the person's "true" probabilities are $p_i$, his expected gain is $\sum p_i \log x_i$ which is maximized when $x_i = p_i$. This elicitation function is not entirely satisfactory, since he may over-estimate the probability of unlikely events to avoid large losses.

There are a number of objective methods of generating probabilities that are sophisticated versions of the principle of insufficient reason—they attempt to give probabilities that correspond to no information, in the hope that any information may be incorporated using Bayes theorem. It is necessary in every case to assume some prespecified probabilities on which to build the "indifferent" probabilities.

We have rested the tortoise four square upon the elephant; but what does the elephant stand on? A method of basing probabilities on similarity judgments is proposed.

## 5.1. Information

If $f$ is a density with respect to $\mathscr{X}$, and $PX = Q(fX)$ each $X$ in $\mathscr{X}$, write $f = dP/dQ$ and $P = Q.f$.

The *information of P with respect to Q* is $I(P, Q) = Q((dP/dQ) \log(dP/dQ))$

defined whenever $(dP/dQ)\log(dP/dQ)\in\mathscr{X}$. If $P$ is a discrete distribution over the integers, and $Q$ gives value $\log 2$ to each integer, $-I(P,Q)$ coincides with Shannon's (1948) definition of entropy, which may be interpreted as the average number of bits per observation, required to send over a channel a stream of observations from $P$ encoded into binary digits. See Good (1966) for a statistical interpretation. If $P$ and $Q$ are both unitary probabilities $I(P,Q)$ is defined by Kullback and Leibler (1951). In this case $I(P,Q)\geqq 0$, and $I(P,Q)=0$ when $P=Q$, so $I(P,Q)$ may be interpreted as a measure of distance between $P$ and $Q$.

For a given $Q$, it is sometimes useful to find the *minimal information $P$* satisfying various constraints—this is put forward as the *principle of maximum entropy* by Jaynes (1957) and Kullback (1959), but there is no reason to think that such a probability is correct for betting—it is merely the probability closest to $Q$ in a certain way. Of course the minimal $P$ always depends on the underlying probability $Q$. See also Christensen (1981).

**Theorem.** *If there exists a density $f = \exp(\sum_{i=1}^{n}\lambda_i X_i)$ such that $Q(fX_i)=a_i$, $i=1,\ldots,n$, then $P$ with $dP/dQ = f$ is uniquely of minimal information among all $P$ satisfying the constraints $P(X_i)=a_i$, $i=1,\ldots,n$.*

PROOF. $g\log f \leqq g\log g + f - g$ with equality only if $f = g$.

If a density $g$ satisfies the constraints, $Q(gX_i)=a_i$,

$$Q(g\log f) = Q[g\sum\lambda_i X_i] = Q[f\sum\lambda_i X_i]$$
$$= Q(f\log f).$$

Thus $Q(f\log f) = Q(g\log f) \leqq Q(g\log g)$ with equality only if $f = g$ as $Q$. Therefore $P$ with $f = dP/dQ$ is of minimal information among all $P$ satisfying the constraints, and no other minimal $P$ exists.

*Note*: Since $I(P,Q)$ is convex as a function of $P$, it may be shown that the minimal information $P$ is unique if it exists.

*Note*: It may happen that no minimal information $P$ exists, but that there exists $P_0 = Q.f_0$ such that for each $A$ with $0 < P_0(A) < \infty$, $P_A = Q(f_0A)/P_0(A)$ is the minimal probability $P$ under the additional constraints $PA = 1$, $P(X) = 0$ if $XA = 0$. The $P_A$'s are conditional probabilities given $A$ corresponding to $P_0$, and $P_0$ may reasonably be taken to be the minimal $P$. For example, if $Q$ is uniform on the integers, the optimal $P$ under the constraints that all but a finite number of probabilities are zero and $P1 = 1$, has all non-zero probabilities equal. Thus the overall minimal $P$ should be taken to be uniform on the integers.

## 5.2. Maximal Learning Probabilities

Let $P$ be a probability on $\mathscr{X}$ and let $X$, $Y$ and $X \times Y$ be random variables defined on $\mathscr{X}$. Suppose that a quotient probability $P_X^Y$ exists, satisfying $P^{X,Y} = P^X P_X^Y$.

If $P^{X,Y}$ has density $f$ with respect to $P^X \times P^Y$ (that is

$$P^{X,Y}W = P^X P^Y[fW] \quad \text{for} \quad W \in \mathscr{X} \times \mathscr{Y}),$$

define *the information between X and Y* to be $P^{X,Y}[\log f] = P^{X,Y}[\log(dP^{X,Y}/dP^X \times P^Y)]$. This measure is zero if $X$ and $Y$ are independent ($P_X^Y = P^Y$), and it is non negative if $P$ is unitary. Define the conditional information of $Y$ given $X$ by $I(Y|X) = I(P_X^Y, P^Y)$; then $P^X[I(Y|X)]$ is the information between $X$ and $Y$. See Lindley (1956).

If $P^X$ is $\sigma$-finite, and $P_X^Y$ has density $h$ with respect to a probability $Q^Y$, $h = dP_X^Y/dQ^Y$, then $P^Y$ has density $P^X h_T$ with respect to $Q$, and

$$P[\log dP/dP^X \times dP^Y] = P^X[I(P_X^Y, Q)] - I[P^Y, Q],$$

which may be interpreted as the probable change of information about $Y$ with respect to $Q$, due to learning $X$.

If $P_X^Y$ is known, $P^{X,Y}$ is determined by $P^X$; Good (1969), Zellner (1977) and Bernardo (1979) have suggested determining $P^X$ to be a "maximal learning probability," maximizing $P^{X,Y} \log(dP^{X,Y}/P^X \times P^Y)$; thus learning $Y$ will maximally increase the information about $X$.

**Theorem.** *Let $P^{X,Y}$ be unitary, with marginal probabilities $P^X$, $P^Y$ and quotient probability $P_X^Y$. For $P_X^Y$ fixed, if there exists $P^X$ with*
 (1)  $P^X\{I(P_X^Y, P^Y) = c\} = 1$
 (2)  $I(P_X^Y, P^Y) \leqq c$
*then the maximal information between X and Y is c, and $Q^X$ produces maximal information if and only if it satisfies (1), (2) and $P^Y = Q^Y$.*

PROOF.     $Q^X[I(P_X^Y, Q^Y) - I(P_X^Y, P^Y)]$
           $= Q^X P_X^Y[\log(dP^Y/dQ^Y)]$
           $= Q^Y \log dP^Y/dQ^Y \leqq 0$ with equality only if $P^Y = Q^Y$.

Thus $Q^X I[P_X^Y, Q^Y] \leqq c = P^X[I(P_X^Y, P^Y)]$ with equality only if $P^Y = Q^Y$ and if (1) and (2) are satisfied by $Q$.                                                                 □

EXAMPLE. Let $Y = 0$ or $1$, let $X$ be a binomial parameter $0 \leqq X \leqq 1$, so that $P_X Y = X, PY = PX$.

$$I[P_X^Y, P^Y] = (1 - X)\log\frac{1 - X}{1 - PX} + X\log\frac{X}{PX}.$$

If $P^X\{0\} = P^X\{1\} = 1/2$, then $I(P_X^Y, P^Y) = \ln 2$ at $X = 0, 1$ and $I[P_X^Y, P^Y] = (1 - X)\log(1 - X) + X\log X - \ln 2 \leqq \ln 2$ for all $X$. Thus $P^X$ is a maximal learning probability, and it is unique.

For two binomial observations, the maximal learning probability is $P^X\{0\} = 15/34$, $P^X\{1/2\} = 4/34$, $P^X\{1\} = 15/34$; empirical evidence suggests that the maximal learning probability is carried by $(n + 1)$ atoms for $n$

observations and converges weakly to the Jeffreys distribution with $\sin^{-1}\sqrt{X}$ uniform.

## 5.3. Invariance

Let $X$ be a random variable from $U, \mathscr{Z}$ to $S, \mathscr{X}$. Let $\sigma$ be a $1-1$ transformation of $S$ onto itself such that $f\sigma \in \mathscr{X}$ each $f \in \mathscr{X}$. A probability $P^X$ is *relatively invariant under* $\sigma$, or $\sigma$-*invariant* if

$$P^X = kP^{\sigma X} \quad \text{some } k, \text{ that is}$$

$$Pf(X) = kPf(\sigma X), \quad \text{each } f \text{ in } \mathscr{X}.$$

(If $P^X$ is unitary, $k = 1$; note that $P^{\sigma X}f = P^X f\sigma$ for *any* $\sigma$.)

For example, let $X = X_1, X_2, \ldots, X_{10}$ denote the results, in heads or tails, of tossing a penny ten times; and suppose we have no reason to differentiate between the tosses—each order is equally likely. Let $\sigma$ be a permutation of 10 numbers; $\sigma X = X_{\sigma 1}, X_{\sigma 2}, \ldots, X_{\sigma 10}$. Then $X$ and $\sigma X$ have the same distribution.

A quotient probability $P^X_Y$ is *relatively invariant under* $\sigma, \tau$ if $\sigma$ is a $1-1$ transformation of $S, \mathscr{X}, \tau$ is a $1-1$ transformation of $T, \mathscr{Y}$ and

$$P^X_Y = kP^{\sigma X}_{\tau Y} \quad \text{some } k.$$

(Note that $P^X_Y f = P^X_{\tau Y}f\tau$ for any $1-1$ $\tau$.) If $P^Y$ is relatively invariant under $\tau$, and $P^X_Y$ is relatively invariant under $\sigma, \tau$, it follows from the product rule that $P^{X,Y}$ is relatively invariant under $\sigma, \tau$—that is $P^{\sigma X, \tau Y} = kP^{X,Y}$. Conversely, if $P^Y$ is relatively invariant under $\tau$ and $P^{X,Y}$ is relatively invariant under $\sigma, \tau$, and if $P^X_Y$ is defined, then $P^X_Y$ is relatively invariant under $\sigma, \tau$ as $P^Y$. More precisely, for each $f \in \mathscr{X}, g \in \mathscr{Y}, P^Y(|P^X_Y f - kP^{\sigma X}_{\tau Y}f|g) = 0$.

Invariances are used to generate prior probabilities as follows. Suppose that $X$ is an observation, $Y$ is an unknown parameter. A model specifies the quotient probability $P^X_Y$, and transformations $\sigma$ and $\tau$ are found such that $P^X_Y$ is $\sigma, \tau$-invariant. It is now assumed that the same invariance applies to the posterior distribution of $Y$ given $X$; this will occur if the prior distribution is $\tau$-invariant. Thus each invariance found on the model induces a constraint on the prior; conceivably, we might find so many invariances on the model that no prior satisfies them all! Here, we are arguing by analogy that observing $\sigma X$ given $\tau Y$ is similar to observing $X$ given $Y$, since the probability models are the same—therefore conclusions about $\tau Y$ given $\sigma X$ should correspond to conclusions about $Y$ given $X$. See Hartigan (1964), Stone (1970), and also Fraser (1968) for a non-Bayesian theory of inference using invariances.

EXAMPLE. Let $X$ and $Y$ be real valued random variables with $(P^X_Y f)(t) = \int f(s)h(s-t)ds$ some density $h$. Equivalently, $X - Y$ has density $h$ given $Y$ with respect to lebesgue measure. Let $\sigma X = X + c, \tau Y = Y + c$.

Then   $P^{\sigma X}_{\tau Y} f = P^X_{\tau Y} f \sigma = P^X_Y (f \sigma) \tau^{-1}$   for any $\sigma, \tau$.

$$P^{\sigma X}_{\tau Y} f(t) = P^X_Y (f\sigma)(t - c) = \int f(s + c)h(s - t + c)ds = P^X_Y f(t).$$

Thus $\sigma, \tau$ is an invariant transformation for $P^X_Y$, and so $P^Y$ is required to be invariant under $\tau$. Thus $P^Y$ has density $P$, $P(t) = e^{\lambda t}$, with respect to lebesgue measure. The posterior density for $P^X_Y$ is $e^{\lambda t}f(s - t)/\int e^{\lambda t}f(s - t)dt$ with respect to lebesgue measure—and $P^Y_X$ is $\tau, \sigma$-invariant.

## 5.4. The Jeffreys Density

Let a distance $\rho$ be a non-negative function on $S \times S$, and assume $\{s \mid \rho(t, s) \leq c\}$ lies in $\mathscr{X}$ on $S$. A local $\rho$-probability on $\mathscr{X}$ is a probability $P$ such that

$$\lim_{c \downarrow 0} P\{\rho(t_1, s) \leq c\}/P\{\rho(t_2, s) \leq c\} = 1   \text{ for each } t_1, t_2 \text{ in } S.$$

Such a probability gives approximately equal value to small spheres.

Jeffreys (1946) considered a number of measures of distance between two probabilities. Write $f = dP/dQ$ if $P$ has density $f$ with respect to Q.

Information distance:      $I(P, Q) = P\left(\log \dfrac{dP}{dQ}\right)$

Root or Hellinger distance:   $r(P, Q) = Q\left[\left(\dfrac{dP}{dQ}\right)^{1/2} - 1\right]^2$

Absolute distance:       $a(P, Q) = Q\left|\dfrac{dP}{dQ} - 1\right| = \sup_{|X| \leq 1} (PX - QX)$.

Since $(u^{1/2} - 1)^2 \leq |u - 1|$, $r(P, Q) \leq a(P, Q)$. Using Schwartz's inequality, for $P$ and $Q$ unitary, $a(P, Q) \leq 2r(P, Q)^{1/2}$. See Pitman (1979) for applications of $r(P, Q)$ to non-Bayesian inference.

**Lemma.** *Let* $\{P_n\}$ *and* $P$ *be unitary. Then* $r(P_n, P) \to 0$ *as* $n \to \infty$ *if and only if* $dP_n/dP \to 1$ *in* $P$.

PROOF.  Let $f_n = dP_n/dP$.

Then   $r(P_n, P) \to 0 \Rightarrow P[f_n^{1/2} - 1]^2 \to 0 \Rightarrow f_n^{1/2} \to 1$ in $P \Rightarrow f_n \to 1$ in $P$.

Conversely

$$f_n \to 1 \text{ in } P \Rightarrow f_n^{1/2} \to 1 \text{ in } P \Rightarrow |f_n^{1/2} - 1|\{|f_n^{1/2} - 1| < \varepsilon\} \to 0 \text{ in } P$$
$$\Rightarrow P|f_n^{1/2} - 1|\{|f_n^{1/2} - 1| < \varepsilon\} \to 0$$
$$\Rightarrow Pf_n^{1/2}\{|f_n^{1/2} - 1|\} < \varepsilon\} \to 1 \text{ since } P\{|f_n^{1/2} - 1| < \varepsilon\} \to 1$$

$$P(f_n^{1/2} - 1)^2 = 2 - 2Pf_n^{1/2} \leq 2 - 2P\{f_n^{1/2}|f_n^{1/2} - 1| < \varepsilon\} \to 0. \qquad \square$$

**Theorem.** *Let $\mathscr{P}$ be a family of unitary probabilities $P_t$ on $\mathscr{X}$, indexed by $T$, a compact subset of $R^p$ such that the interior of $T$ is dense in $T$. Assume*

(1) *$P_t$ has density $f_t$ with respect to some probability $\mu$ on $\mathscr{X}$.*

(2) *$f_t = f_s$ as $\mu \Rightarrow s = t$.*

(3) *for all $t$, there exists a vector derivative $(\partial/\partial t)f_t^{1/2}$ such that*

$$h(s, t) = \left| f_s^{1/2} - f_t^{1/2} - (s - t)' \frac{\partial}{\partial t} f_t^{1/2} \right| \bigg/ |s - t| \to 0 \text{ as } |s - t| \to 0$$

*where $|s - t|$ is euclidean distance.*

(4) *For $|s - t| < \delta_t > 0$, $h(s, t) < Z_t$ where $\mu(Z_t^2) < \infty$.*

*The probability $J$ that has density with respect to lebesgue measure on $T$ equal to the determinant $j(t) = |\mu[(\partial/\partial t)f^{1/2}((\partial/\partial t)f^{1/2})']|^{1/2}$, is a local $\rho$-measure where $\rho$ is the Hellinger distance, provided $j(t)$ is continuous and non-zero in $T$.*

PROOF. $r(P_s, P_t) = \mu(f_s^{1/2} - f_t^{1/2})^2$. Fix $t$.

As $r(P_s, P_t) \to 0, f_s/f_t \to 1$ in $P_t$; let $s_0$ be a limit point of the sequence of $s$ values (by compactness, $s_0$ exists). From $(3), f_s \to f_{s_0}$. Therefore $f_{s_0}/f_t = 1$ in $P_t$ which implies $s_0 = t$ from (2).

Thus $r(P_s, P_t) \to 0$ if and only if $s \to t$, so that the set of $s$ values with $r(P_s, P_t) < \varepsilon$ may be found in a neighbourhood of $t$.

From (3), $r(P_s, P_t) = (s - t)'\mu[(\partial/\partial t)f_t^{1/2}((\partial/\partial t)f_t^{1/2})'](s - t) + o(|s - t|^2)$.

The sphere $r(P_s, P_t) \leqq \varepsilon$ corresponds to an approximate ellipsoid in $T$, $(s - t)'\sum_t(s - t) \leqq \varepsilon$ of volume $K\varepsilon^{1/2p}|\sum_t|^{-1/2}$ where $j(t) = |\sum_t|^{1/2}$. The probability of $r(P_s, P_t) \leqq \varepsilon$, with the specified density $j(t)$, is $K\varepsilon^{1/2p}[1 + o(1)]$ since $j$ is continuous and positive. The density $j(t)$ thus generates a local $\rho$-probability. $\square$

The Jeffreys density was put forward simultaneously by Jeffreys (1946) and by Perks (1947). Perks considered confidence regions for $t$ which may be constructed from a sequence of independent observations each distributed as $P_t$. The confidence region in the neighbourhood of $t$ has volume asymptotically proportional to $j_t^{-1}$, under certain regularity conditions similar to those in the theorem. Thus if $t_0$ is the true value of $t$, we will have a confidence region closely concentrated near $t_0$ if $j_t$ is large; Perks places density $j_t$ on $t$ to represent this expectation. A more explicit confidence justification is given by Welch and Peers (1966), for the case where $T$ is the real line: after $n$ observations from $\mathscr{X}$, the conditional probability of $t$ given $n$ observations is $P_n$; choose $t_{n,\alpha}$ so that $P_n(t < t_{n,\alpha}) = \alpha$; under regularity conditions the confidence size of the interval estimate $\{t < t_{n,\alpha}\}$ is $P_t(t < t_{n,\alpha}) = \alpha + 0(n^{-1/2})$, but for Jeffreys' density $P_t(t < t_{n,\alpha}) = \alpha + 0(n^{-1})$. Thus the Jeffreys density gives Bayes one-sided intervals which are more nearly confidence intervals than the intervals for any other prior. It should be noted that the same justification does not hold for two-sided intervals, Hartigan (1966).

The Jeffreys density is also obtained from maximum learning probabilities (Bernardo (1979): suppose that $n$ independent observations with probability $P_t$ generate the space $\mathscr{X}_1 \times \ldots \mathscr{X}_n$, and let $\mathscr{Y}$ be the Baire functions on $T$. As $n \to \infty$, the information between $\mathscr{X}_1 \times \ldots \mathscr{X}_n$ and $\mathscr{Y}$, denoted by $I_n$, satisfies, under regularity conditions

$$I_n - \tfrac{1}{2} \log n \to -I(P^{\mathscr{Y}}, J) + K.$$

Thus the maximal learning probability for the asymptotic information between $\mathscr{X}_1 \times \ldots \mathscr{X}_n$ and $\mathscr{Y}$, is the Jeffreys probability $J$.)

The Jeffreys probability is induced on the indexing set $T$ by the family of probabilities $\mathscr{P} = \{P_t, t \in T\}$; it will provide the same probability on $\mathscr{P}$ regardless of the particular set $T$ used to index $\mathscr{P}$. The topology on $T$ is induced by the Hellinger distance on $\mathscr{P}$. The probability on $T$ is unchanged if Jeffreys' probability is computed using a number of observations from $\mathscr{X}$ rather than a single observation. These properties are also possessed by the family of densities $P_\alpha(t)$

$$\frac{d}{dt} \log p_\alpha(t) = \left\{ P_t \left( \frac{\partial}{\partial t} \log f \frac{\partial^2}{\partial t^2} \log f \right) + \alpha P \left( \frac{\partial}{\partial t} \log f \right)^3 \right\} \Big/ P \left( \frac{\partial}{\partial t} \log f \right)^2$$

for $t$ one dimensional, Hartigan (1965). This family gives the Jeffreys probability when $\alpha = 1/2$, and often generates commonly accepted prior densities with suitable choice of $\alpha$. Perhaps there is an interpretation in differential geometry.

If a subset of $\mathscr{P}$ is considered, $\mathscr{P}' = \{P_t, t \in T'\}$, where $T'$ is a compact set in $R^k$ whose interior is dense in $T'$, then the Jeffreys probability on $T'$ may be obtained by conditioning the probability on $T$ to $T'$, provided $J(T') > 0$. If however the indexing set $T$ is partitioned into a family $\{T'_\alpha\}$ of indexing sets of lower dimensionality, the Jeffreys probabilities on each of the $T'_\alpha$ might not be conditional probabilities from Jeffreys' probability on $T$; the Jeffreys construction is not consistent with the combination of conditional probabilities.

## 5.5. Similarity Probability

Let $X, Y$ and $X \times Y$ be random variables from some probability space into $(S, \mathscr{X}), (T, \mathscr{Y})$ and $(S \times T, \mathscr{X} \times \mathscr{Y})$. If $P^{X,Y}$ has density $l$ with respect to $P^X \times P^Y$ (that is $P^{X,Y} W = P^X P^Y [lW]$ for $W$ in $\mathscr{X} \times \mathscr{Y}$), call $l$, a real valued function on $S \times T$, the *likeness* or *similarity* between $S$ and $T$.

The random variable $Y$ describes a number of possible outcomes in the past, the random variable $X$ describes outcomes in the future, and $l$ specifies similarities between pairs of these outcomes. We propose that $l$ be specified subjectively to correspond to perceived similarities, and that the probabilities $P^{X,Y}$, $P^X$ and $P^Y$ be determined from $l$. (Hartigan, 1971). In the notation of 5.2, $l = dP_Y^X/dP^X$, the density of the quotient probability $P_Y^X$ relative to the marginal probability $P^X$. If $X$ and $Y$ are both discrete,

$$l(x, y) = P(X = x, Y = y)/P(X = x)P(Y = y).$$

If $P^X$ and $P^Y$ have densities with respect to some probabilities $Q^X$ and $Q^Y$

$$l(x, y) = p^{X,Y}(x, y)/p^X(x)p^Y(y)$$

where $p^{X,Y}$, $p^X$ and $p^Y$ are densities of $P^{X,Y}$, $P^X$ and $P^Y$.

EXAMPLE 1: *Selecting from a deck of cards.* A deck of cards is composed of 52 cardboard rectangles of apparently identical dimensions, one side of the rectangles being distinguished by different markings, the other side marked the same for all cards. The deck is shuffled with the uniform side showing. What is the probability that the top card is the ace of spades?

Let $X$ denote the top card. Let $Y$ denote past knowledge about this deck of cards, observations of the shuffling process, and any other information. For $x$ one of the expected 52 cards, and for $y$ past knowledge that does not refer to a particular card, take $l(x, y)$ to be constant. Thus

$$P(X = x, Y = y) = cP(X = x)P(Y = y)$$
$$P(X = x|Y = y) = cP(X = x).$$

Now consider the event that the top card is either $x$, one of the 52, or a card with a picture of a rabbit (the children have been at the cards again). Call this event $\{X = x \text{ or } R\}$.

Then

$$P[X = x \text{ or } R|Y = y] = c'P[X = x \text{ or } R], \qquad c' \neq c$$
$$P[X = R|Y = y] = c'P[X = R] + (c' - c)P[X = x].$$

Thus $P(X = x|Y = y)$ and $P(X = x)$ are the same for all $x$. I do not feel happy about the rabbit, but some event of different similarity is necessary to show that all probabilities are equal. Note that $P[X = x|Y = y]$ is the same for all $x$ only if the knowledge $y$ contains nothing to distinguish the cards.

This looks like the principle of insufficient reason, but it is not subject to partitioning paradoxes. Consider for example the rotatable cards—21 cards that look different when rotated through $180°$. Distinguish between the two versions of these cards when selecting the top card, so that there are 73 possible results. There are now a number of different similarities.

$l(x, y)$: rotatable $x$ to a typical $y$
$l(z, y)$: non-rotatable $z$ to a typical $y$
$l(x \text{ or } R, y)$: rotatable $x$ or rabbit to a typical $y$
$l(z \text{ or } R, y)$: non-rotatable $z$ or rabbit to a typical $y$
$l(x' \text{ or } x, y)$: either version of a rotatable $x$ to a typical $y$.

Assume that $l(x'$ or $x, y) = l(z, y)$ but that the other three similarities are differ-
ent from each other and from $l(z, y)$. Then $P(x|y) = P(x'|y)$ and $P(x$ or $x'|y) =$
$P(z|y)$ so that the probabilities are 1/52 for non-rotatable cards and 1/104
for rotatable cards.

EXAMPLE 2: *Uniform on the integers.* It is not possible to present realistic
examples of infinite sample spaces in a bounded universe, but such sample
spaces have proved to be mathematically convenient. Who would give up
Poisson and normal distributions?

Let $X$ be a random variable taking integer values.

Let $Y$ be past knowledge.

Suppose $y$ is such that no integer for $X$ is preferred, and take $l(x, y)$ to be
the same for all $x$. Again we need some outside event $E$ such that $l(x$ or $E, y)$
is the same for all $x$, and different to $l(x, y)$. Then $P[X = x|y]$ is the same for
all $x$, and the distribution on the integers is uniform. The uniform distribution
on the line may be handled similarly—it will require that, $l(\{x_1 \leqq x \leqq x_2\}, y)$
depend only on the length of the interval $\{x_1 \leqq x \leqq x_2\}$.

Nothing much is happening, just the transfer of equal similarity perceptions
to equal probability statements.

EXAMPLE 3: *A sequence of coin tosses.* Let $X_1, X_2, X_3, \ldots$ denote the
sequence of heads or tails in tossing a coin. Let $Y$ be past knowledge about
this and other coins and other things. If $x = x_1, x_2, \ldots, x_n$ is a particular
sequences of $n$ tosses, let $x' = x_{\sigma 1}, x_{\sigma 2}, \ldots, x_{\sigma n}$ denote a permutation of $x$.
Suppose $l(x, y) = l(x', y)$ for all permutations $x'$, and suppose

$$l(x_1, y) \neq l(x_1 \text{ or } x_2, y) \text{ for } x_2 \text{ not a permutation of } x_1.$$

$$l(x_1 \text{ or } x_2, y) = l(x'_1 \text{ or } x'_2, y).$$

Then $P(x|y) = P(x'|y)$ and the sequence $X_1, X_2, \ldots$ is exchangeable.

The probability distribution for $X_1, X_2, \ldots$ is then independent Bernoulli
given $p = \lim(\sum x_i/n)$, which exists almost surely. Frequency theory assumes
no more than this—thus the small probability assumptions of frequency
theory may be derived from equal similarities of permuted sequences to
given knowledge. To have a full probability model for a sequence of coin
tosses, it is necessary to specify in addition the prior distribution of $P$, that
is to specify the similarities of various $p$ values to given knowledge.

Assume that only a finite number of $p$ values are possible, say $p_1, p_2, \ldots, p_N$.
Then

$$\frac{P(p_i|y)}{P(p_j|y)} = \frac{1 - \dfrac{l(p_i \text{ or } p_j, y)}{l(p_j, y)}}{\dfrac{l(p_i \text{ or } p_j, y)}{l(p_i, y)} - 1}$$

More generally, $P_y(p \in A)/l(p \in A, y) = P_y[\{p \in A\}/l(p, y)]$.

If the distribution of $p$ given $y$ has a density $f_y(p)$ with respect to lebesgue measure, differentiation of this formula gives

$$\frac{f_y(p_0)}{l(p \leq p_0, y)} + \left( \int_0^{p_0} f_y(p) dp \right) \frac{d}{dp_0} \left[ \frac{1}{l(p \leq p_0, y)} \right] = \frac{f_y(p_0)}{l(p_0, y)}.$$

For example, if $l(p \leq p_0, y) = p_0$    for   $0 \leq p_0 \leq 1/2$,

$$l(p_0, y) = 2p_0 \quad \text{for} \quad 0 \leq p_0 \leq 1/2,$$
$$f_y(p_0) = cp_0 \quad \text{for} \quad 0 \leq p_0 \leq 1/2.$$

We might, by symmetrical similarity judgments, require that $f_y(p)$ be symmetrical about $p = 1/2$.

I might be charged with replacing mystical methods of determining priors by mystical methods of specifying similarities. I am not proposing formal methods of determining similarities. They are subjective judgments relating expected events to past knowledge; they may come only as comparative judgments—this $p$ value is more similar than that; even such comparative judgments may usefully constrain the conditional distribution given $y$.

## 5.6. Problems

E1. Find the minimum information probability with respect to lebesgue measure on the plane, with means, variances and covariance fixed.

E2. Let $X > 0$ have minimum information with respect to lebesgue measure subject to $P(X) = 1$, and let $Y > 0$ have minimum information with respect to lebesgue measure subject to $P(Y^2) = 1$. Show that $X$ and $Y^2$ have different distributions, illustrating lack of invariance of minimum information probabilities.

E3. Let $X$ and $Y$ be two integer variables with fixed marginal distributions. Find the minimum information joint distribution of $X$ and $Y$ with respect to uniform probability on pairs of integers $(i, j)$, $-\infty < i < \infty$, $-\infty < j < \infty$.

P1. A person ranks $N$ candy bars, which have delectability coefficients $d_1, d_2, \ldots, d_N$ such that the $i$th bar is preferred to the $j$th bar with probability $d_i/(d_i + d_j)$. Find the minimum information probability for the complete ranking of the candy bars, with the probabilities that $i$ is ranked above $j$ as given, with respect to uniform probability over permutations of the candy bars.

P2. Let $X_1, \ldots, X_n$ be $n$ discrete variables with known pairwise distributions. Find the minimum information probability for the joint distribution of $X_1, \ldots, X_n$ with respect to counting measure on atoms.

Q1. Show that a minimal information $P$ with respect to $Q$ may exist, satisfying $n$ constraints $PX_i = a_i$, but not satisfying

$$dP/dQ = \exp\left( \lambda_0 + \sum_{i=1}^n \lambda_i X_i \right) \quad \text{for any } \lambda_0, \lambda_i.$$

P3. Let $\mathcal{Y}, \mathcal{X}_1, \ldots, \mathcal{X}_n, \ldots$ be probability subspaces of $\mathcal{X}$ such that $\mathcal{X}_n \uparrow$. Assume that a conditional probability $P_n : \mathcal{X} \to \mathcal{X}_n$ exists for each $n$ with $P_n P_{n+1} = P_n$. Assume $I(\mathcal{Y} | \mathcal{X}_n) \in \mathcal{X}_n$. Then $I(\mathcal{Y} | \mathcal{X}_n)$ is a sub-martingale, that is

$$P_{n-1}[I(\mathcal{Y} | \mathcal{X}_n)] \geq I[\mathcal{Y} | \mathcal{X}_{n-1}].$$

(Roughly translated, we expect to learn something by knowing $\mathcal{X}_n$.)

P4. Let $P_\mu^X$ be a normal distribution on $\mathcal{X}$ with mean $\mu$, variance 1. Find the invariant transformations $(\sigma, \tau)$ for the quotient probabilities $P_\mu^X$, and show that the only probability on $\mu$ which is $\tau$-invariant for every $\tau$ is lebesgue measure.

P5. Let $P_V^X$ be a normal distribution on $\mathcal{X}$ with mean 0, variance $V$. Find the invariant transformations $(\sigma, \tau)$ for $P_V$, and find measures on $V$ which are $\tau$-invariant (more than one!).

P6. Let $P_V^X$ be a bivariate normal distribution on $\mathcal{X}$, with means 0 and covariance matrix $V$. Find the invariant transformations $(\sigma, \tau)$ for $P_V$, and find measures on $V$ which are $\tau$-invariant.

E4. In the binomial case, find that function of $p$ which is uniformly distributed according to the Jeffreys density.

E5. Compute the Jeffreys density in the bivariate normal case with unknown means and covariances.

P7. A contingency table has 1000 cells with cell probabilities $p_1, \ldots, p_{1000}$, with $\sum_{i=1}^{1000} p_i = 1$. Show that the Jeffreys density implies that the number of cells with $p_i > 1/1000$ is approximately $N(160, 134)$. Suppose that examination of the contingency table produced 12 empirical frequencies greater than 1/1000. Would you use the Jeffreys density for constructing estimates of $p_i$?

P8. An observation comes from the normal mixture, $N(\theta_1, 1)$ with probability $\frac{1}{2}$ and $N(\theta_2, 1)$ with probability $\frac{1}{2}$. Find the Jeffreys probability for $(\theta_1, \theta_2)$.

P9. Observe the toss of a coin, with unknown success probability $p$, until $r$ successes appear. Find the Jeffreys density for $p$. Now observe $n$ tosses of the coin, and find the Jeffreys density for $p$. You observe a man toss a coin 50 times, getting 20 successes, and he asks you, as consulting Bayesian, to compute the posterior density of $p$. In an effort to be impartial, you do so with the Jeffreys density for $p$ for 50 tosses of a coin. He then confides in you that he stopped tossing when 20 successes were reached. Do you change the posterior density?

P10. Let $P_\theta$ have density $(1 + \theta' x)/4\pi$ with respect to uniform probability on the three dimensional sphere, for each $x, \theta$ in the sphere. Show that the Jeffreys probability is uniform over the sphere. If $\theta$ is constrained to lie in a great circle through the poles, show that Jeffreys probability is uniform over the great circle. Show that the constrained Jeffreys probability is not a conditional probability for the unconstrained Jeffreys probability. (Similarly it is not possible to have joint probabilities and conditional probabilities which are rotation invariant.)

# 5.7. References

Bernardo, J. M. (1979), Reference posterior distributions for Bayesian inference (with discussion), *J. Roy. Statist. Soc.* **41**, 113–147.

Christensen, Ronald (1981), *Entropy Minimax Sourcebook, Vol. I: General Description.* Lincoln, Massachusetts: Entropy Limited.

Fraser, D. A. S. (1968), *The Structure of Inference.* New York: John Wiley.

Good, I. J. (1966), A derivation of the probabilistic explanation of information, *J. Roy. Statist. Soc.* B **28**, 578–581.

Good, I. J. (1969), What is the use of a distribution?, in Krishnaiah (ed.), *Multivariate Analysis* Vol. II, 183–203. New York: Academic Press.

Hartigan, J. A. (1964), Invariant prior distributions, *Ann. Math. Statist.* **35**, 836–845.

Hartigan, J. A. (1965), The asymptotically unbiased prior distribution, *Ann. Math. Statist.* **36**, 1137–1152.

Hartigan, J. A. (1966), Note on the confidence-prior of Welch and Peers, *J. Roy. Statist. Soc.* B **28**, 55–56.

Hartigan, J. A. (1971), Similarity and probability, in V. P. Godambe and D. A. Sprott, (eds.), *Foundations of Statistical Inference.* Toronto: Holt, Rinehart and Winston.

Jaynes, E. T. (1957), Information theory and statistical mechanics, *Phys. Rev.* **106**, 620–630.

Jeffreys, H. (1946), An invariant form for the prior probability in estimation problems, *Proc. R. Soc. London* A **186**, 453–461.

Kullback, S. (1959), *Information Theory and Statistics.* New York: Wiley.

Kullback, S. and Leibler, R. A. (1951), On information and sufficiency, *Ann. Math. Statist.* **22**, 79–86.

Lindley, D. V. (1956), On a measure of the information provided by an experiment, *Ann. Math. Statist.* **27**, 986–1005.

Perks, W. (1947), Some observations on inverse probability; including a new indifference rule, *J. Inst. Actuaries* **73**, 285–334.

Pitman, E. J. G. (1979), *Some Basic Theory for Statistical Inference.* London: Chapman and Hall.

Shannon, C. E. (1948), A mathematical theory of communication, *Bell System Tech. J.* **27**, 379–423.

Stone, M. (1970), Necessary and sufficient conditions for convergence in probability to invariant posterior distributions, *Ann. Math. Statist.* **41**, 1939–1953.

Zellner, A. (1977), Maximal data information prior distributions, in A. Aykac and C. Brumat, (eds.), *New Developments in the Applications of Bayesian Methods*, p. 211–232. Amsterdam: North Holland.

Welch, B. L. and Peers, H. W. (1963), On formulae for confidence points based on integrals of weighted likelihoods, *J. Roy. Statist. Soc.* B **25**, 318–329.

Winkler, R. L. (1967), The assessment of prior distributions in Bayesian analysis, *J. Am. Stat. Assoc.* **62**, 776–800.

# CHAPTER 6

# Decision Theory

## 6.0. Introduction

Fisher (1922) compared two estimators by considering their distributions given an unknown parameter of interest. For example, in estimating a normal distribution mean the sample mean is unbiased with variance $2/\pi$ times the variance of the sample median, for all values of the distribution mean, so it is to be preferred to the sample median. Of course, it may be difficult in general to decide between the two families of distributions.

Neyman and Pearson (1933) proposed evaluation of a test statistic by considering the probability of rejection of the null hypothesis under various values of the parameter of interest.

Wald (1939) proposed a general theory to cover both of these cases, in which a general decision function (of the data) is evaluated by its average loss for each value of the parameter. Wald suggested minimax techniques for selecting decisions, that arose out of von Neumann's theory of games— we play a game against nature (the unknown parameter value) so that our loss will not be too severe if nature chooses the worst parameter value.

Ramsey, de Finetti and Savage use similar ideas from the theory of games in showing that a coherent betting strategy requires a probability distribution on the set of bets. There is no technical or conceptual difference between coherent betting and admissible decision making. If we decide to use one decision function rather than another, we are accepting the bet corresponding to the difference in losses for the two decision functions.

## 6.1. Admissible Decisions

It is necessary to choose one of a set of decisions $D$. The consequences of the decisions are determined by the outcome $s$ in a set of possible outcomes $S$. For decision $d$, and outcome $s$, there is a loss $L(d, s)$. Since there is no reason to differentiate between two decisions which have the same loss for each value of $s$, one may regard the decision set $D$ as a set of real valued functions on $S : d(s)$ is the loss incurred by decision $d$ if outcome $s$ occurs.

Let $d_1 \leqq d_2$ mean $d_1(s) \leqq d_2(s)$ for $s \in S$. Say that $d$ is *admissible* if it is minimal in $D$, that is $d' \leqq d, d' \in D \Rightarrow d' = d$. A *complete class* $C$ is a subset of $D$ such that $d' \in D - C$ implies $d \leqq d'$ for some $d$ in $C$. A complete class is minimal if it contains no proper complete class. It is easy to show that if a minimal complete class exists, it is the set of admissible decision functions. However, no admissible decision functions may exist; consider $S = \{1\}$, $D = \{d | -\infty < d(s) < \infty\}$; every decision is inadmissible, no minimal complete class exists. See Wald (1939, 1950) for the first theory, and Ferguson (1967) and Berger (1980) for expository texts.

If $P$ is a probability on a probability space $\mathscr{X}$ on $S$, such that $D \subset \mathscr{X}$, a decision $d_0$ is a *Bayes decision* if $P(d_0) \leqq P(d)$, $d \in D$. If a Bayes decision is unique it must be admissible (for otherwise there exists $d' \leqq d_0, d' \neq d_0$, which implies $P(d') \leqq P(d_0)$, so that $d'$ is also a Bayes decision, contradicting uniqueness). We will say that $d_0$ is $P$-Bayes.

If $P$ is a finitely additive probability on $\mathscr{X}$, a linear space of functions including $D$, then $d_0$ is $P$-Bayes if $P(d - d_0) \geqq 0$, $d \in D$.

**Theorem.** *Let $P$ be a finitely additive probability on $\mathscr{X}$ including $D$. If $d_0$ is unique $P$-Bayes, then $d_0$ is admissible. If $d_0$ is admissible and $D$ is a convex space of bounded functions then $d_0$ is $P$-Bayes with respect to a finitely additive probability $P$ on a space $\mathscr{X}$ including $D$ [Heath and Sudderth (1978)].*

PROOF. If $d_0$ is $P$-Bayes, then $P(d - d_0) \geqq 0, d \in D$. If $d \leqq d_0$, then $P(d - d_0) \leqq 0$, so $d$ must be $P$-Bayes. Since $d_0$ is unique, $d = d_0$ and so $d_0$ is admissible.

If $d_0$ is admissible, let $\mathscr{X}$ be the bounded real valued functions, and define $\mathscr{P} = \{X | d_0 + aX \geqq d, \text{ some } a > 0, \text{ some } d \in D\}$.

Then $X_1, X_2 \in \mathscr{P} \Rightarrow aX_1 + bX_2 \in \mathscr{P}$ for $a \geqq 0, b \geqq 0$ by convexity of $D$. From Theorem 2.1, it is possible to extend $\mathscr{P}$ to $\mathscr{P}^* \supset \mathscr{P}$, such that $\mathscr{P}^* \cap -\mathscr{P}^* = \mathscr{P} \cap -\mathscr{P}$ and $\mathscr{P}^* \cup -\mathscr{P}^* = \mathscr{X}$. Note that $\mathscr{P}$ includes all $X \geqq 0$, and excludes all $X < 0$.

Set $PX = \sup \{\alpha | X - \alpha d_0 \in \mathscr{P}^*\}$. Then $-\infty < PX < \infty$ because $X$ is bounded, and $P$ is an additive functional on $X$ with $PX \geqq 0$ for $X \in \mathscr{P}$; in particular $P$ is non-negative, that is, $PX \geqq 0$ for $X \geqq 0$.

Since $d_0 + (d - d_0) \geqq d$, $d - d_0 \in \mathscr{P}$ and $P(d - d_0) \geqq 0$ all $d \in D$. Thus $d_0$ is $P$-Bayes as required. $\square$

*Note.* The idea of this theorem is that accepting $d_0$ over all other decisions $d$ is accepting the bets $d - d_0$, all $d \in D$. You will surely lose money (that is, $d_0$ is inadmissible) unless there is a finitely additive probability $P$ such that $P(d - d_0) \geqq 0$ all $d \in D$.

Apparently we should be satisfied with finitely additive probability; however finitely additive probabilities do not discriminate well between decisions, so that a unique $P$-Bayes decision is unusual; many inadmissible decisions may be also optimal for a given functional $P$.

Consider for example estimation of a normal mean. The data is $x$, the unknown mean $\theta$, and the decision is a function of $x$, say $\delta$. In the loss framework, the decision will be represented as a real valued function of $\theta$, $d(\theta) = \int (\delta(x) - \theta)^2 \exp[-\frac{1}{2}(x - \theta)^2]dx/\sqrt{2\pi}$. Consider $D$ to be composed of decisions $d$: estimate $\theta$ by $\delta_i(x)$ with probability $\alpha_i$, where $\delta_i(x) - x \to 0$ as $|x| \to \infty$, $i = 1, \ldots, n$. For all such decisions $d(\theta) \to 1$ as $|\theta| \to \infty$. Uniform finitely additive probability on $\theta$ gives value $\lim_{|\theta| \to \infty} X(\theta)$ to $X$ when this limit exists. Thus all the decisions proposed are finitely additive Bayes with respect to the uniform distribution; they are not all admissible, demonstrating that optimality by a finitely additive probability is rather too easy. See Heath and Sudderth (1972, 1978).

## 6.2. Conditional Bayes Decisions

Let $X$ be a random variable into $S$, $\mathcal{X}$ and let $Y$ be a random variable into $T$, $\mathcal{Y}$ and assume that $X \times Y$ is a random variable into $S \times T$, $\mathcal{X} \times \mathcal{Y}$. Assume there is a quotient probability $P_X^Y$ on $Y$ given $X$, and denote the value of $P_X^Y$ at $s$ by $P_X^Y(s)$; this defines a probability on $\mathcal{Y}$. Similarly assume a quotient probability $P_Y^X$. By the product rule $PZ = P^X P_X^Y Z = P^Y P_Y^X Z$ each $Z$ in $\mathcal{X} \times \mathcal{Y}$.

Now suppose a decision $d$ in $D$ is to be taken using an observation $t$ in $T$. The loss, if $d$ is taken when the parameter $s$ is true, is $L(d, s)$. A family $\Delta$ of decision functions $\delta : T \to D$ is constructed satisfying $(s, t) \to L(\delta(t), s) \in \mathcal{X} \times \mathcal{Y}$ for each $\delta \in \Delta$. The loss associated with $\delta$ is the risk $r(\delta, s) = P_X^Y(s)[L(\delta, s)]$.

**Theorem.** *Suppose that, for each $t$, $\delta_0(t)$ is $P_Y^X(t)$-Bayes. If $\delta_0 \in \Delta$, then $\delta_0$ is $P^X$-Bayes. Conversely if $\delta_0^*$ is $P^X$-Bayes, and $\delta_0(t)$ is $P_Y^X(t)$-Bayes, then $\delta_0^*(t)$ is $P_Y^X$-Bayes as $P^Y$.*

PROOF.   $P_Y^X(t)[L(\delta_0(t), \cdot) - L(d, \cdot)] \leqq 0$   all $d$.

$\qquad P_Y^X(t)[L(\delta_0(t), \cdot) - L(\delta(t), \cdot)] \leqq 0$   all $\delta \in \Delta$.

Since    $P^X P_Y^X = P^Y P_X^Y$,

$\qquad P^X P_Y^X[L(\delta_0, \cdot) - L(\delta, \cdot)] \leqq 0$   all $\delta \in \Delta$

$\qquad P^X[r(\delta_0, \cdot) - r(\delta, \cdot)] \leqq 0$   all $\delta \in \Delta$.

Thus $\delta_0$ is $P^X$-Bayes.

Conversely if $\delta_0^*$ is also $P^X$-Bayes,

$$P^Y P_Y^X(t)[L(\delta_0(t), \cdot) - L(\delta_0^*(t), \cdot)] \geqq 0$$
$$P_Y^X(t)[L(\delta_0(t), \cdot) - L(\delta_0^*(t), \cdot)] = 0 \text{ as } P^Y. \qquad \square$$

*Note.* The theorem makes it practicable to find Bayes decision functions, since it is easier to search over the smaller space of decisions $D$ to obtain a conditional Bayes decision, than to search over the larger space of decision functions $\Delta$.

EXAMPLE. Let $S$, $T$ be the real line, let $P_X^Y(s)$ denote the normal distribution with mean $s$ and variance 1. Let $\mathscr{X}$ and $\mathscr{Y}$ be the Baire functions on $S$ and $T$. Take $P^X$ to be normal with mean 0 and variance $\sigma^2$. Then $P_Y^X(t)$ is normal with mean $\sigma^2 t/(1 + \sigma^2)$ and variance $\sigma^2/(1 + \sigma^2)$.

Let $D$ be the real line, $L(d, s) = (d - s)^2$, and let $\Delta$ be the set of Baire functions on $D$. Then $r(\delta, s) = \int (\delta(t) - s)^2 \exp[-\frac{1}{2}(t - s)^2] dt / \sqrt{2\pi}$. For a particular $t$, the $P^X(t)$-Bayes decision $\delta_0(t)$ minimizes

$$\int (d - s)^2 \exp\left[ -\frac{1}{2}\left( s - \frac{\sigma^2 t}{1 + \sigma^2} \right)^2 \frac{1 + \sigma^2}{\sigma^2} \right] ds, \delta_0(t) = \frac{\sigma^2 t}{1 + \sigma^2}.$$

Thus the $P^X$-Bayes decision function is $\delta_0(t) = \sigma^2 t/(1 + \sigma^2)$.

It may happen that $\delta_0$ is conditionally Bayes (that is $\delta_0(t)$ is $P_Y^X(t)$-Bayes for each $t$), but not Bayes because $\delta_0 \notin \Delta$. In the present example, if $P^X$ is uniform, $\delta_0(t) = t$ is the conditional Bayes decision but it has risk $r(\delta_0, s) = 1$ which is not integrable, so it is not the Bayes decision; in certain cases, conditional Bayes decisions are even inadmissible (see Chapter 9 on many means). It should not be thought that the possible inadmissibility of conditionally Bayes decisions is caused by $P^X$ not being unitary; in the present example, if $P^X = N(0, \sigma^2)$, and $L(d, s) = \exp(\frac{1}{2}s^2/\sigma^2)(d - s)^2$, the conditional Bayes estimate is $\delta_0(t) = t$, but it is not the Bayes estimate. In the same way, the inadmissible estimate for many normal means is conditionally Bayes with respect to a unitary prior distribution. See also Chapter 7 on conditional bets.

## 6.3. Admissibility of Bayes Decisions

If a Bayes decision $d_0$ is unique it is admissible [for any decision beating it would also have to be the Bayes decision].

If the decisions $d$ in $D$ are continuous in some topology on $S$, and the finitely additive probability $P$ is supported by $S$ (that is, $Pf > 0$ if $f$ is continuous, non-negative, and not identically zero), then any decision which is $P$-Bayes is admissible.

[If $d_0$ is $P$-optimal and $d' \leq d_0$, then $P(d_0) \leq P(d') \Rightarrow P(d_0 - d') = 0$. Thus $d' = d_0$ since $P$ is carried by $S$.]

Let $P$ be a probability on $\mathscr{X}$ on $S$. Say that $P$ is *supported by $S$* if for each continuous $X$ in $\mathscr{X}$, $X \neq 0$, $X \geq 0$ implies $PX > 0$. Say that $P$ is $X_n$-$\sigma$-finite if some sequence $X_n$ in $\mathscr{X}$ has $X_n \uparrow 1$. Say that a decision $d_0$ is $X_n$-limit Bayes if $\sup_{d \in D} P[X_n(d_0 - d)] \to 0$ as $n \to \infty$.

**Theorem.** *If P is supported by S and is $X_n$-$\sigma$-finite for some continuous $X_n$, and if D consists of continuous functions, and if $d_0$ is $X_n$-limit Bayes, then $d_0$ is admissible.*

PROOF. If $d' \leq d_0$ and $d' \neq d_0$, $X_n(d_0 - d') \geq 0$, $X_n(d_0 - d') \neq 0$ for $n$ large enough, so $P[X_n(d_0 - d')] > 0$ for $n$ large. Since $X_n(d_0 - d') \uparrow (d_0 - d')$, $P[X_n(d_0 - d')] \to 0$ is impossible. Thus $d_0$ is admissible.                $\square$

*Note.* If $P$ is carried by $S$, if $D$ consists of continuous functions in $\mathscr{X}$, and if $d_0$ is $P$-Bayes, then $d_0$ is admissible. The present theorem applies to decisions $d_0$ which may not lie in $\mathscr{X}$.

EXAMPLE. Let $x$ be an observation from $N(\theta, 1)$, and suppose that $\theta$ is to be estimated with squared error loss function. The theorem will be used to show that $x$ is admissible, being $X_n$-limit Bayes with respect to uniform probability $\mu$ on $\theta$.

The estimate $\delta$ generates the decision

$$d(\theta) = \int (\delta(x) - \theta)^2 \exp[-\tfrac{1}{2}(x - \theta)^2] dx/\sqrt{2\pi}$$

The decisions $d$ are $\theta$-continuous. The measure $\mu$ is $\sigma$-finite with respect to $f_n(\theta) = \exp(-\theta^2/n) \uparrow 1$ as $n \to \infty$

$$\inf_d \mu(f_n d) = \inf_\delta \int\int (\delta(x) - \theta)^2 \exp[-\tfrac{1}{2}(x - \theta)^2 - \theta^2/n] dx d\theta/\sqrt{2\pi}$$

$$= \int\int \inf_\delta (\delta - \theta)^2 \exp[-\tfrac{1}{2}(x - \theta)^2 - \theta^2/n] d\theta dx/\sqrt{2\pi}$$

$$= \int \exp[x^2/(n + 2)] dx \left/ \left(1 + \frac{2}{n}\right)^{3/2}\right.$$

$$= \sqrt{\pi} n^{3/2}/(2 + n)$$

$$\mu(f_n) = \sqrt{n\pi}$$

$$\sup_d \mu[f_n(1 - d)] = \sqrt{n\pi}\left[1 - \frac{n}{2 + n}\right] = 2\sqrt{n\pi}/(n + 2) \to 0.$$

Since the estimate $\delta_0(x) = x$ generates the decision $d \equiv 1$, $\delta_0$ is continuous $\sigma$-finite Bayes as required. (Here the decision $d \equiv 1$ does not lie in $\mathscr{X}$.)

EXAMPLE 2. Let $r$ be a binomial observation, with $P_p\{r\} = \binom{n}{r} p^r (1 - p)^{n-r}$,

$0 \leq r \leq n$. Suppose that $p$ is to be estimated with squared error loss function— the estimator $\delta$ corresponds to

$$d(p) = \sum_{r=0}^{n} (\delta(r) - p)^2 \binom{n}{r} p^r (1 - p)^{n-r}.$$

For the measure $\mu$, $\mu(f) = \int_0^1 f(p)[p(1-p)]^{-1} dp$, take the functions

$$f_\alpha(p) = [p(1-p)]^\alpha \uparrow 1 \text{ as } \alpha \downarrow 0.$$

$$\inf_d \mu(df_\alpha) = \inf_\delta \int \sum_{r=0}^n (\delta(r) - p)^2 \binom{n}{r} p^r(1-p)^{n-r} p^{\alpha-1}(1-p)^{\alpha-1} dp$$

$$= \sum_{r=0}^n \inf_\delta \int (\delta - p)^2 \binom{n}{r} p^r(1-p)^{n-r}[p(1-p)]^{\alpha-1} dp$$

$$= \sum_{r=0}^n \binom{n}{r} \frac{\Gamma(r+\alpha)\Gamma(n-r+\alpha)}{\Gamma(n+2\alpha)} \cdot \frac{(r+\alpha)(n-r+\alpha)}{(n+2\alpha)^2(n+2\alpha+1)}$$

$$= \sum_{r=0}^n \binom{n}{r} \frac{\Gamma(r+\alpha+1)\,\Gamma(n-r+\alpha+1)}{\Gamma(n+2\alpha+2)\,(n+2\alpha)} \to \frac{1}{n} \text{ as } \alpha \to 0.$$

For $\delta_0 = \dfrac{r}{n}$, $\mu(d_0 f_\alpha) = \int \dfrac{p(1-p)}{n} p^{\alpha-1}(1-p)^{\alpha-1} dp$

$$= \int p^\alpha(1-p)^\alpha dp/n = \Gamma^2(\alpha+1)/[\Gamma(2\alpha+2)n]$$

$$\to \frac{1}{n} \text{ as } \alpha \to 0.$$

Thus $\sup_d \mu[f_\alpha(d_0 - d)] \to 0$ as $\alpha \to 0$. Also $\mu$ is carried by the interval $[0, 1]$. Therefore $d_0$ is admissible.

## 6.4. Variations on the Definition of Admissibility

A decision $d$ is beaten by $d'$ at $\theta$ if $d(\theta) > d(\theta')$. We say

$d$ is *somewhere beaten* by $d'$ if $d'(\theta) \leq d(\theta)$ all $\theta$, $d'(\theta) < d(\theta)$ some $\theta$,
$d$ is *everywhere beaten* by $d'$ if $d'(\theta) < d(\theta)$ all $\theta$
$d$ is *uniformly beaten* by $d'$ if $\inf[d(\theta) - d'(\theta)] > 0$.

Then $d$ is *admissible* in a set of decisions $D$ if $d$ is not somewhere beaten by any $d'$ in $D$. Say that $d$ is *weakly admissible* if it is not everywhere beaten by any $d'$ in $D$, and that $d$ is *very weakly admissible* if it is not uniformly beaten by any $d'$ in $D$.

The sense of admissibility appropriate for finitely additive probabilities is *very weak*. If $d_0$ is a finitely additive Bayes decision with respect to $P$, then $d_0$ is very weakly admissible. [Heath and Sudderth (1978).] Conversely, the argument of Theorem 6.1 shows that if $d_0$ is very weakly admissible, and $D$ is convex with $\sup_s d(s) < \infty$ each $d$ in $D$, then $d_0$ is finitely additive Bayes with respect to some $P$.

The sense of admissibility appropriate for probabilities is *weak*. If $d_0$ is a Bayes decision with respect to $P$, then $d_0$ is weakly admissible. However, converse results are more complicated than in the finitely additive case. See for example Farrell (1968).

If $D$ consists of continuous functions and $S$ is compact, then a finitely additive $P$ on the space of continuous functions on $S$ is countably additive (since if a decreasing sequence of functions converges to zero it converges uniformly to zero). Thus if $d_0$ is weakly admissible, it is $P$-Bayes with respect to a unitary probability $P$. More generally if $D$ consists of continuous functions zero outside compact subsets of $S$, a weakly admissible $d_0$ is $P$-Bayes with respect to a unitary probability $P$. [If $d_0$ is carried by $S'$, consider decisions and probabilities restricted to $S'$.]

## 6.5. Problems

E1. Let the sample space $S$ be finite. Let $D$ be a set of decisions on $S$ (real valued functions on $S$). Let $P$ give positive probability to each non-zero non-negative $X$ on $S$. Show that a $P$-Bayes decision is admissible.

E2. For decisions $D$ on a finite $S$, show that no $P$-Bayes decision may exist, and that it might not be admissible if it does exist.

P1. Let $X$ be a Poisson observation with $P(X = x) = \lambda^x e^{-\lambda}/x!$. Show that $X$ is an admissible estimate of $\lambda$ with squared error loss.

P2. Let $X$ be binomial with $P(X = x) = \binom{n}{x} p^x (1 - p)^{n-x}$, $0 \leqq x \leqq n$. Consider estimates $\delta$ of $p$ using squared error loss. Show that the estimate $\delta(x) = x/n$ is weakly admissible.

## 6.6. References

Berger, James O. (1980), *Statistical Decision Theory*. New York: Springer-Verlag.

Farrell, R. (1968), Towards a theory of generalized Bayes tests, *Ann. Math. Statist.* **39**, 1–22.

Ferguson, T. S. (1967), *Mathematical Statistics, a Decision Theoretic Approach*. New York: Academic Press.

Fisher, R. A. (1922), On the mathematical foundations of theoretical statistics, *Phil. Trans. Roy. Soc.* A **222**, 309–368.

Heath, D. C. and Sudderth, W. D. (1972), On a theorem of de Finetti, odds making, and game theory, *Ann. Math. Statist.* **43**, 2072–2077.

Heath, D. C. and Sudderth, W. D. (1978), On finitely additive priors, coherence, and extended admissibility, *Ann. Statist.* **6**, 333–345.

Neyman, J. and Pearson, E. S. (1933), On the problem of the most efficient tests of statistical hypotheses. *Phil. Trans. Roy. Soc.* A **231**, 289–337.

Neyman, J. and Pearson, E. S. (1933), The testing of statistical hypotheses in relation to probabilities a priori, *Proc. Camb. Phil. Soc.* **24**, 492–510.

Wald, A. (1939), Contributions to the theory of statistical estimation and testing hypotheses, *Ann. Math. Statist.* **10**, 299–326.

Wald, A. (1950), *Statistical Decision Functions*. New York: John Wiley.

# CHAPTER 7

# Uniformity Criteria for Selecting Decisions

## 7.0. Introduction

The set of admissible decision functions in a particular problem is usually so large that further criteria must be introduced to guide selection of a decision function. Many such criteria require that the unknown parameter values be treated "uniformly" in some way; decision procedures are required to be invariant or unbiased or minimax or to have confidence properties. Since selection of a decision function, from a Bayesian point of view, is selection of a probability distribution on the parameter values according to which the decision function is optimal, these criteria may be viewed as methods of selecting indifference probability distributions on the parameter values.

The general conclusion is that the various uniformity criteria are satisfied by no unitary Bayes decision procedures, establishing the necessity for considering non-unitary probabilities.

## 7.1. Bayes Estimates Are Biased or Exact

Let $P$ be a probability on $\mathcal{X}$, let $\theta \in \mathcal{X}$ be such that a conditional probability $P[X|\theta]$ exists satisfying $PX = P[P(X|\theta)]$ for all $X$. Let $\mathcal{Y}$ be a subspace of $\mathcal{X}$.

An estimate $Y$ in $\mathcal{Y}$ of $\theta$ is *unbiased* if $P[Y|\theta] = \theta$ and *exact* if $P[Y \neq \theta] = 0$. A Bayes estimate of $\theta$ in $\mathcal{Y}$ with respect to squared error loss is a $Y$ such that $P[Y - \theta]^2$ is a minimum over $P[Y^* - \theta]^2$ with $Y^* \in \mathcal{Y}$, $(Y^* - \theta)^2 \in \mathcal{X}$.

**Theorem.** *An unbiased Bayes estimate is exact.*

PROOF. Let $Y$ be an unbiased Bayes estimate. Define $X^A = (-A) \vee X \wedge A$ for each $A \geq 0$. Note $|X|A \geq (X^A)^2 \in \mathcal{X}$.

63

Setting $Y^* = Y + \varepsilon Y^A$,

$$P[Y - \theta]^2 \leq P[Y^* - \theta]^2 = P[Y - \theta]^2 + 2\varepsilon P[Y^A(Y - \theta)] + \varepsilon^2 P(Y^A)^2.$$

Taking $\varepsilon$ small and of sign $P[Y^A(\theta - Y)]$,

$$PY^A[\theta - Y] = 0.$$

If $Y$ is unbiased,

$$P[Y|\theta] = \theta$$
$$P[Y - \theta|\theta] = 0$$
$$P[(Y - \theta)\theta^A|\theta] = 0$$
$$P[(Y - \theta)\theta^A] = 0.$$

Thus if $Y$ is Bayes and unbiased,

$$P[Y - \theta][\theta^A - Y^A] = 0.$$

Since $x - y \geq x^A - y^A$ if $x \geq y$,

$$P[\theta^A - Y^A]^2 \leq 0, \quad \text{which implies} \quad P[\theta^A \neq Y^A] = 0 \quad \text{all } A,$$

$P[\theta \neq Y] = 0$, so $Y$ is exact.

*Note.* It may happen that a posterior mean be unbiased. For example, let $Y$ given $\theta$ be distributed as $N(\theta, 1)$ and let $\theta$ be uniform on the line; the posterior mean of $\theta$ given $Y$ is $Y$, and

$$P[[\theta - Y]^2|Y] \leq P[[\theta - \delta(Y)]^2|Y] \quad \text{for all borel functions } \delta.$$

Also $Y$ is unbiased, $P[Y|\theta] = \theta$. However $Y$ is not the Bayes estimate of $\theta$ in the class of functions $\delta(Y)$ because none of the functions $[\delta(Y) - \theta]^2$ are integrable.

## 7.2. Unbiased Location Estimates

Let $X_1, X_2, \ldots, X_n$ and $\theta$ be real valued random variables on $\mathscr{X}$. Suppose $X_1, \ldots, X_n$ are independent and identically distributed given $\theta$, and that $X_1 - \theta$ given $\theta$ has a distribution which does not depend on $\theta$; assume that this distribution has density $f$ with respect to Lebesgue measure.

An *invariant estimator* $\delta$ of $\theta$ satisfies $\delta(\mathbf{X} + a) = \delta(\mathbf{X}) + a$.

**Theorem.** *Suppose that $X_1$ has finite second moment given $\theta$. The Pitman estimator, the posterior mean of $\theta$ given $\mathbf{X}$ corresponding to a uniform prior probability on $\theta$, is unbiased and has minimum mean square error given $\theta$ of all invariant estimators.*

PROOF. Consider first the case of one observation. Any invariant estimator is of the form $\delta(X) = X + a$ and has mean square error var $X + [P_\theta \delta(X) - \theta]^2$, so the Pitman estimate $\delta_0$ will be optimal if it is unbiased.

Now

$$\delta_0 = \int \theta f(x - \theta) d\theta / \int f(x - \theta) d\theta = \int (x - u) f(u) du$$
$$= x - \int u f(u) du$$
$$P_\theta \delta_0 = \int x f(x - \theta) dx - \int u f(u) du = \theta.$$

Thus the Pitman estimator is unbiased and optimal.

For $n$ observations, consider the behavior of an invariant estimator $\delta(X)$ and the Pitman estimator $\delta_0(X)$ conditional on $X_2 - X_1, X_3 - X_1, \ldots,$ $X_n - X_1$. The conditional density of $X_1$ is $\prod f(X_i - \theta) / \int \prod f(X_i - \theta) d\theta$; the conditional Pitman estimate corresponding to this density is just $\delta_0(X)$.

Also $\delta(X)$ is an invariant estimator of $\theta$, considered as a function of $X_1$ with $X_i - X_1$ fixed, $i = 2, \ldots, n$. Thus the conditional risk of $\delta_0$ is no greater than that of $\delta$, and hence the unconditional risk of $\delta_0$ is optimal. Similarly, since $\delta_0$ is conditionally unbiased for $\theta$, it is unconditionally unbiased. □

*Note*: The Pitman estimator is not the Bayes estimator corresponding to a uniform prior, because it has constant risk which is not integrable. Stein (1959) shows that the Pitman estimator is admissible whenever $X_1$ has finite third moment given $\theta$, and Brown and Fox (1974) have shown admissibility under weak conditions.

## 7.3. Unbiased Bayes Tests

In testing, a decision is made whether a parameter $\theta$ lies in a set $H_0$ or in a set $H_1$, $H_0 H_1 = 0$, $H_0 + H_1 = 1$. Thus $d$ takes the values $H_0$ or $H_1$ and has loss

$$L(d, \theta) = \alpha \{d = H_1\} H_0 + \beta \{d = H_0\} H_1.$$

The loss is $\alpha$ if you mistakenly decide $\theta \in H_1$ and $\beta$ if you mistakenly decide $\theta \in H_0$.

A decision function based on a random variable $Y$ is a function $\delta(Y)$ taking values $H_0$ or $H_1$. The decision function is *unbiased* if $P_\theta L(\delta(Y), \theta) \leq P_\theta L(\delta(Y), \theta')$ for $\theta, \theta'$, which is equivalent to

$$P_\theta[\delta = H_1] \leq \beta/(\alpha + \beta) \leq P_{\theta'}[\delta = H_1] \quad \text{for } \theta \in H_0, \theta' \in H_1.$$

Since $\alpha$ and $\beta$ are usually arbitrary, the more usual definition of unbiasedness requires

$$P_\theta[\delta = H_1] \leq P_{\theta'}[\delta = H_1] \quad \text{for } \theta \in H_0, \theta' \in H_1.$$

Suppose now that $Y$ and $\theta$ are random variables on $\mathscr{X}$, a probability $P$ is defined on $\mathscr{X}$, and a quotient probability $P_Y^\theta$ exists such that $P^\theta P_\theta^Y = P^Y P_Y^\theta$. The conditional Bayes decision $\delta(Y)$ minimizes $P_Y[\alpha\{\delta(Y) = H_1\}H_0 + \beta\{\delta(Y) = H_0\}H_1]$ which requires

$$P_Y[\theta \in H_1] \leq \beta/(\alpha + \beta) \leq P_{Y'}[\theta \in H_1] \quad \text{for } \delta(Y) = H_0, \delta(Y') = H_1.$$

Compare the form of the Bayes decision with the unbiasedness requirement. The conditional Bayes decision $\delta$ is the Bayes decision if $L(\delta(Y), \theta) \in \mathscr{X}$. The Bayes decision is saying the obvious, that you decide $\theta \in H_0$ if the conditional probability of $H_0$ is large, and that you decide $\theta \in H_1$ if the conditional probability of $H_0$ is small.

To test $\theta = \theta_0$ against $\theta \neq \theta_0$, assume that $P_\theta^Y$ has density $f_\theta(Y)$ with respect to some probability $Q$. The posterior distribution of $\theta$ given $Y$ is given by

$$(P_Y^\theta g)(t) = P[g(\theta)f_\theta(t)]/P[f_\theta(t)]$$

and the conditional Bayes decision is:

$$\delta(t) = 1 \text{ if } P\{\theta = \theta_0\}f_{\theta_0}(t)/P[f_\theta(t)] \leq \beta/(\alpha + \beta).$$

For a given prior $P$ on $\theta$, consider the mixture $P_a$,

$$P_a X = aX(\theta_0) + PX.$$

The conditional Bayes decision is $\delta(t) = 1$ if $f_{\theta_0}(t)/P[f_\theta(t)] \leq k$ where $k$ depends on $a$, $\alpha$, $\beta$. The atom at $\{\theta_0\}$ affects only $k$. A test of this form will be called a *P-Bayes test for $\theta = \theta_0$ against $\theta \neq \theta_0$*.

**Theorem.** *Let $\theta$ be a real valued random variable. Let $Y$ be a random variable such that $P_\theta^Y$ has density $f_\theta$ with respect to a probability $Q$. Assume that $f_\theta(t)$ is $\theta$-differentiable for each $t$, and $\sup_{\theta \in I}|(df_\theta)/d\theta|$ is $Q$-integrable for each finite interval $I$ of $\theta$ values. Let the prior probability for $\theta$ be $P^\theta$.*

*The test $\delta(t) = 1$ if $f_{\theta_0}(t)/P[f_\theta(t)] \leq k$ is unbiased for every $\theta_0$, $k$ if and only if $f_{\theta_0}(t)/P[f_\theta(t)]$ has the same distribution for every $\theta_0$, letting $t$ have the distribution of $Y$.*

PROOF. Let $h(t) = P^\theta f_\theta(t)$, $g(\theta, Y) = f_\theta(Y)/h(Y)$. Unbiasedness requires

$$P_\theta^Y\{g(\theta, \cdot) \leq k\} \leq P_{\theta'}^Y\{g(\theta, \cdot) \leq k\}$$
$$Q\{g(\theta, \cdot) \leq k\}(f_\theta - f_{\theta'}) \leq 0$$
$$Q\left[\frac{df_\theta}{d\theta}\{g(\theta, \cdot) \leq k\}\right] = 0$$

where the differentiation is justified because $\sup_{\theta \in I}|(df_\theta)/d\theta|$ is $Q$-integrable.

$$Q\left[h\frac{dg(\theta, \cdot)}{d\theta}\{g(\theta, \cdot) \leq k\}\right] = 0$$

$$Q\left[h\frac{dg(\theta,\cdot)}{d\theta}\phi[g(\theta,\cdot)]\right]=0 \quad \text{for bounded continuous } \phi$$

$$\frac{d}{d\theta}Q[h\phi(g(\theta,\cdot))]=0 \quad \text{for } \phi \text{ twice differentiable.}$$

But $P(\phi[g(\theta,\cdot)])=Q[h\phi(g(\theta,\cdot))]$ for $\theta$ fixed. Thus $g(\theta,Y)$ must have the same distribution for all $\theta$. The converse follows by running the steps of the proof in reverse. $\qquad\square$

*Note.* Unbiased Bayes tests for unitary $P$ rarely exist, but the above condition is met for some other $P$. For example, if $Y \sim N(\theta,1)$ and $\theta$ is uniform, the Bayes test is accept $\theta = \theta_0$ if $|Y-\theta_0| \leq k$, and the test statistic $|Y-\theta_0|$ has the same distribution for all $\theta_0$ since the distribution of $Y$ is uniform.

## 7.4. Confidence Regions

Let $Y$ and $\theta$ be random variables, let $P_\theta^Y$ be the quotient probability on $Y$ given $\theta$. Suppose we wish to select a set of likely $\theta$-values; a decision $d$ will be a set of $\theta$-values, and a decision function $\delta(Y)$ selects such a set for each $Y$ value.

A set selection function $\delta$ is a *confidence procedure* if $P_\theta^Y[\theta \in \delta] = \alpha_0$ for all $\theta$. This requirement is analogous to invariance or unbiasedness in that all $\theta$-values are given the same treatment.

Consider a family of testing decisions $d(\theta)$, where $d(\theta_0) = 1$ decides $\theta = \theta_0$ and $d(\theta_0) = 0$ decides $\theta \neq \theta_0$. The set $\{d(\theta) = 1\}$ is selected by $d$, giving a correspondence between families of tests and set selection decisions.

The loss (analogous to testing loss) for $d$ is

$$L(d,\theta_0,\theta) = \alpha\{d(\theta_0) \neq 1\}\{\theta = \theta_0\} + \beta\{d(\theta_0) = 1\}\{\theta \neq \theta_0\}$$

$$P_\theta^Y[L(\delta,\theta_0,\theta)] = \alpha P_\theta^Y[\theta_0 \notin \delta]\{\theta = \theta_0\} + \beta P_\theta^Y[\theta_0 \in \delta]\{\theta \neq \theta_0\}$$

Thus we want a large probability $P_\theta^Y[\theta \in d]$ and a small probability $P_\theta^Y[\theta_0 \in d]$ with $\theta \neq \theta_0$. The standard decision theory is not applicable because of the appearance of both $\theta$ and $\theta_0$ in the loss function; it is necessary to consider a prior distribution over $\theta$ and $\theta_0$ to discover admissible set selection procedures $\theta$. For a given prior $P^{\theta,\theta_0}$, the Bayes set $d$ given $Y$ minimizes

$$\alpha P_Y^{\theta,\theta_0}\{d(\theta_0) \neq 1\}\{\theta = \theta_0\} + \beta P_Y^{\theta,\theta_0}\{d(\theta_0) = 1\}\{\theta \neq \theta_0\}$$

which requires $d(\theta_0) = 1$ if $P_{Y,\theta_0}^\theta\{\theta = \theta_0\} \geq \beta/(\alpha+\beta)$.

For a given prior $P^\theta$ on $\theta$, the conditional prior $P_{\theta_0}^\theta$ suggested by testing is

$$P_{\theta_0}^\theta X = aX(\theta_0) + P^\theta X,$$

and then the set selection procedure is $d(\theta_0) = 1$ if

$$f_{\theta_0}(Y)/P^\theta[f_\theta(Y)] \leq K$$

where $Y$ has density $f_\theta$ given $\theta$. Regions of this form are called Bayes high density regions; see for example Box and Tiao (1973) and Hartigan (1966). From theorem 5.2, it follows that unitary Bayes high density regions are confidence regions for all $K$ only if the probability $P$ is a maximal learning probability; in the many cases where maximal learning probabilities do not exist, confidence regions cannot be unitary Bayes high density regions. However, confidence regions are often Bayes high density regions corresponding to non-unitary prior measures. Hartigan (1966) shows that high density regions are asymptotically closest to confidence regions for the Jeffreys density.

## 7.5. One-Sided Confidence Intervals Are Not Unitary Bayes

Let $Y$ be a random variable on $\mathcal{X}$, and let $\theta$ be a real valued random variable on $\mathcal{X}$. A *one-sided confidence interval* $[-\infty, \theta(Y)]$ is such that $P_\theta[\theta \leq \theta(Y)] = \alpha$ all $\theta$. A one-sided Bayes interval $[-\infty, \theta(Y)]$ is such that $P_Y[\theta \leq \theta(Y)] = \alpha$ all $Y$.

**Theorem.** *A one-sided unitary Bayes interval of size* $\alpha$, $0 < \alpha < 1$, *is not a confidence interval.*

Proof. Let $P_Y[\theta \leq \theta(Y)] = \alpha$. Then $P[\theta \leq \theta(Y)] = \alpha$. For each fixed $\theta_0$, $P[\theta \leq \theta(Y) | \theta \leq \theta_0] > \alpha$ if $P[\theta(Y) \geq \theta_0 | \theta \leq \theta_0] > 0$.

If $P[\theta(Y) \geq \theta_0 | \theta \leq \theta_0] = 0$ for all $\theta_0$, $P[\theta(Y) \leq \theta] = 1$ which contradicts $0 < \alpha < 1$. Thus $P[\theta \leq \theta(Y) | \theta \leq \theta_0] > \alpha$ for some $\theta_0$.

If $\theta \leq \theta(Y)$ is a confidence interval, $P_\theta[\theta \leq \theta(Y)] = \alpha$ all $\theta$, so $P_\theta[\theta \leq \theta(Y) | \theta \leq \theta_0] = \alpha$, $P[\theta \leq \theta(Y) | \theta \leq \theta_0] = \alpha$ which is a contradiction. $\square$

## 7.6. Conditional Bets

Let $Y$ and $\theta$ be random variables on $\mathcal{X}$. A bet $Z(Y, \theta)$ is *conditionally probable given $Y$* if $Z(Y, \theta) \in \mathcal{X}$ each $Y$ and $P_Y Z(Y, \theta) \geq 0$. A bet $Z(Y, \theta)$ is *conditional given $Y$* if $P_\theta[Z(Y, \theta) f(Y)] < 0$ all $\theta$ for no $f \geq 0$ such that $Z(Y, \theta) f(Y) \in \mathcal{X}$ all $\theta$.

If $Y$ and $\theta$ take finitely many values, the two conditions are equivalent—for any matrix $X_{ij}$ there exists no $\beta_j \geq 0$ such that $\sum_j X_{ij} \beta_j < 0$ all $i$ if and only if there exists $\alpha_i \geq 0$ ($\alpha \neq 0$) such that $\sum_j X_{ij} \alpha_i \geq 0$. (Equivalently, a convex set disjoint from the negative quadrant is separated from the negative quadrant by a hyperplane.)

In general, the two senses of conditionality are not equivalent. For example, suppose that $Y$ and $\theta$ take values on the integers $1, 2, \ldots$.

Define $P_\theta^Y\{Y = i\}Z(i, \theta) = [-\{\theta \leq i\} + \{\theta > i\}]/i^2$

Then $\sum P_\theta^Y\{Y = i\}Z(i, \theta)g(i) = -\sum_{\theta \leq i}g(i)/i^2 + \sum_{\theta > i}g(i)/i^2$.

This quantity is defined only if $\sum g(i)/i^2$ converges, and thus it cannot be negative for every $\theta$ when $g \geq 0$, and $Z(i, \theta)$ is a conditional bet. For any probability $P^\theta$, $\sum_\theta P^\theta P_\theta^Y\{Y = i\}Z(i, \theta) = [-P(\theta \leq i) + P(\theta > i)]/i^2$ is necessarily negative for some $t$, so $Z$ cannot be conditionally probable.

**Theorem.** *Let $Y$ and $\theta$ be real valued random variables, and suppose that $(-\infty, Y)$ is a confidence interval of size $\alpha$ for $\theta$, and that $0 < P_\theta(Y \leq a) < 1$ for all $a$. Then $Z(Y, \theta) = \{\theta \leq Y\} - \alpha$ is not a conditional bet given $Y$.*

Proof. For $\theta \leq a$,

$$P_\theta(Z(Y, \theta)\{Y \leq a\}) = P_\theta\{\theta \leq Y \leq a\} - \alpha P\{Y \leq a\}$$
$$= (1 - \alpha)P_\theta\{Y \leq a\} - P_\theta\{Y < \theta\}$$
$$= (1 - \alpha)[P_\theta\{Y \leq a\} - 1] < 0.$$

For $\theta > a$,

$$P_\theta(Z(Y, \theta)\{Y \leq a\}) = -\alpha P_\theta(Y \leq a) < 0.$$

Thus $Z(Y, \theta)$ is not a conditional bet. □

*Note.* See Olshen (1973) for references and an application to confidence ellipsoids. From 7.5, we know that one sided confidence intervals are not conditionally probable with respect to a unitary probability, but they may be conditionally probable with respect to a non-unitary probability. If the definition of a conditional bet is weakened so that $Z(Y, \theta)$ is weakly conditional given $Y$ provided $P_\theta^Y[Z(Y, \theta)f(Y)] < 0$ for no $f$ such that $Z(Y, \theta)f(Y) \in \mathscr{X}$ all $\theta$, AND $Z(Y, \theta)f(Y) \in \mathscr{X}$ taking $Y$ and $\theta$ random, then if $Z(Y, \theta)$ is conditionally probable it is weakly conditional given $Y$. Thus for example if $Y \sim N(\theta, 1)$ where $\theta$ is uniform, $\theta < Y + 1.64$ is a 95% confidence interval, conditionally probable, and weakly conditional given $Y$, but not conditional given $Y$. The bets $(\{\theta < Y + 1.64\} - .95)\{Y \leq 0\}$ have negative conditional probability given $\theta$, but are not integrable overall.

Freedman and Purves (1969) and Dawid and Stone (1972) show, under regularity conditions, that the notions of conditionally probable and conditional bet coincide if and only if the distributions $P_Y^\theta$ and $P_\theta^Y$ are constructed according to Bayes theorem.

## 7.7. Problems

E1. Let $t, 0 \leq t \leq n$ be an observation from the binomial distribution with $P_p\{t\} = \binom{n}{t}p^t(1 - p)^{n-t}$. Show that the posterior mean for $p$, corresponding to any prior unitary probability $P$, is biased.

P1. Show that the Bayes estimate $\tilde{\theta}$ corresponding to the loss function $L(d, \theta) = |d - \theta|$ is the median of the posterior probability of $\theta$ given $Y$. Does there exist a non-atomic posterior probability for which $\tilde{\theta}$ is median unbiased, that is

$$P_\theta[\theta < \tilde{\theta}] = P_\theta[\theta \leq \tilde{\theta}] = \tfrac{1}{2} \quad \text{all } \theta?$$

Q1. Let $Y_1, Y_2, \ldots Y_n$ denote independent observations from $f(\theta, Y)$, and let $P$ have density $g(\theta)$ with respect to lebesgue measure on the line. Under suitable regularity conditions, when $\theta_0$ is true, show the posterior mean

$$P[\theta \,|\, Y_1, \ldots, Y_n] = \theta_0 + \frac{1}{n}\left[ -\frac{\partial}{\partial\theta_0} \log g + P_{\theta_0}(f_1 f_2 + f_3)/P_{\theta_0} f_2 \right]\Big/ P_{\theta_0} f_2$$
$$+ 0(n^{-3/2})$$

where $f_i = [\partial^i/\partial\theta^i \log f]_{\theta=\theta_0}$. Then show that $g$ is asymptotically unbiased if

$$g(\theta) = -P_\theta\left[ \frac{\partial}{\partial\theta^2} \log f(\theta, Y) \right]$$

(the square! of the Jeffreys density). Hartigan (1965).

E2. Let $Y, 0 \leq Y < 2\pi$, have density $f(\theta, Y) = \tfrac{1}{2}[1 + \cos(Y - \theta)]$, where $0 \leq \theta \leq 2\pi$. Let the prior probability be uniform over $0 \leq \theta \leq 2\pi$. Show that the Bayes high density region is a confidence region.

Q2. As the number of observations $Y_1, \ldots, Y_n$ becomes infinite, under suitable regularity conditions, find an asymptotic expression for the confidence size of the Bayes high density region with respect to $P$,

$$P_{\theta_0}\{\pi f(\theta_0, Y_i)/P_\theta[\pi f(\theta, Y_i)] > k\}.$$

P2. Let the decision $d$ be an ordering of the parameter values $\theta$, with $L(d, \theta) = l\{d(\cdot) > d(\theta)\} - l\{d(\cdot) < d(\theta)\}$ for some loss measure $l$. Show that the Bayes decision given the observation $Y$ orders $\theta$ according to $dP_Y/dl$.

P3. Let the decision $d$ be an interval $\{-\infty, c\}$ on the line. Let $\theta$ be on the line, and set $L(d, \theta) = -\{\theta \leq c\} + K(c - \theta)^+$. Show that the Bayes decision corresponding to a probability $P$ with density $g$, satisfies

$$K \int_{-\infty}^{c} g(\theta)d\theta = g(c).$$

P4. Consider the 95% confidence interval for $\theta$ based on one sample from $N(\theta, 1)$, $\{\theta \leq Y + 1.64\}$. Bet \$95 to win \$5 that $\theta \leq Y + 1.64$ whenever $Y \geq 0$, and bet \$5 to win \$95 that $\theta > Y + 1.64$ whenever $Y < 0$. Find your probable gain as a function of $\theta$.

P5. In the normal location case, show the 95% confidence interval for $\theta$, $(Y - 1.96, Y + 1.96)$, is not a conditional bet. (Bet an amount proportional to $e^Y$ that $\theta$ lies outside the interval.)

P6. Let $x_1, x_2, \ldots, x_n$ be a sample from $N(\mu, \sigma^2)$, let $\bar{x}$ be the mean of $x_1, \ldots, x_n$ and let $s$ be the standard deviation. Show that the confidence interval for $\mu$, $\bar{x} - ks, \bar{x} + ks)$ can be beaten by betting that the interval contains $\mu$ if $s > 1$, and that it doesn't contain $\mu$ when $s \leq 1$. (Buehler and Fedderson (1963)).

P7. If $y, x_1, \ldots, x_n$ are sampled from $N(\mu, 1)$, show that the 95% tolerance interval for $y$, $\{y < \bar{x} + 1.64[1 + (1/n)]^{1/2}\}$ may be beaten by betting differently according to the value of $\bar{x}$.

E3. The decision $d$ chooses one of two parameter values $s_1, s_2$ and $L(d, s) = |s - d|$. Show that the Bayes decision, given an observation $t$ with density $f(s, t)$, for any prior probability which has $P\{s_1\} > 0$, $P\{s_2\} > 0$, is

$$d = s_1 \quad \text{if} \quad f(s_1, t)/f(s_2, t) > c,$$
$$d = s_2 \quad \text{if} \quad f(s_1, t)/f(s_2, t) \leqq c.$$

P8. $x_1, x_2$ are observations from $N(\mu, \sigma^2)$. A test for $\mu = 0$ against $\mu \neq 0$ is *similar* if the probability of deciding $\mu \neq 0$ when $\mu = 0$ is independent of $\sigma^2$. Are any Bayes tests similar?

P9. If $X$ and $Y$ are random variables with the same distribution, show that $P(X - Y > a) \leqq P(|X| > \frac{1}{2}a)$. Let $P_\theta$ be a family of probability distributions with positive densities $f(\theta, Y)$, $0 \leqq \theta \leqq \infty$, satisfying the conditions of theorem 7.3, such that $f(\theta, Y)/f(\theta_0, Y) \to 0$ as $\theta \to \infty$ for each $Y$. Show that no unbiased unitary Bayes test exists.

P10. Let $x$ be an observation with density $f(x - \theta)$ with respect to lebesgue measure, where

$$f(u) = (2/|u| - 1)f(2 - |u|)\{|u| \leqq 2\}.$$

Let $g$ be a prior density with respect to lebesgue measure,

$$g(\theta) = \{[2\theta] = 2[\theta]\}$$

where $[\theta]$ is the largest integer $\leqq \theta$. Show that the posterior mean with respect to $g$ is unbiased for $\theta$. [The uniform distribution is not the only unbiased distribution in location problems].

## 7.8. References

Box, G. E. P. and Tiao, G. C. (1973), *Bayesian Inference in Statistical Analysis*. Reading: Addison-Wesley.

Buehler, R. J. and Fedderson, A. P. (1963), Note on conditional property of student's t, *Ann. Math. Statist.* **34**, 1098–1100.

Brown, L. D. and Fox, M. (1974), Admissibility in statistical problems involving a location or scale parameter, *Ann. of Statistics* **2**, 807–814.

Dawid, A. P. and Stone, M. (1972), Expectation consistency of inverse probability distributions. *Biometrika* **59**, 486–489.

Freedman, D. and Purves, R. A. (1969), Bayes method for bookies, *Ann. Math. Statist.* **40**, 1177–1186.

Hartigan, J. A. (1965), The asymptotically unbiased prior distribution, *Ann. Math. Statist.* **36**, 1137–1154.

Hartigan, J. A. (1966), Estimation by ranking parameters, *J. Roy. Statist. Soc.* B **28**, 32–44.

Olshen, R. A. (1973), The conditional level of the F-test, *J. Am. Stat. Ass.* **68**, 692–698.

Pitman, E. J. C. (1939), Location and scale parameters, *Biometrika* **30**, 391–421.

Stein, C. (1959), The admissibility of Pitman's estimator of a single location parameter, *Am. Math. Statist.* **30**, 970–979.

# CHAPTER 8
# Exponential Families

## 8.0. Introduction

Let $\mu$ be a probability on $\mathcal{Y}$, and choose a unitary probability $P$ on $\mathcal{Y}$ to minimize the information $P(\log(dP/d\mu))$ subject to $PY_i = c_i, i = 1, \ldots, k$. The optimal probability $P$ has density

$$\frac{dP}{d\mu} = \exp\left[ \sum_{i=1}^{k} a_i Y_i(t) + b \right].$$

Such a $P$ is said to be *exponential* with respect to $\mu$, for the functions $\mathbf{Y}$ and the parameters $\mathbf{a}$, denoted $E[\mu, \mathbf{Y}, \mathbf{a}]$. The further parameter $b$ is determined as a function of $\mathbf{a}$ by $P1 = 1$. An *exponential family* $\{P_\mathbf{s}, \mathbf{s} \in S\}$ consists of $E[\mu, \mathbf{Y}, \mathbf{s}]$, $\mathbf{s} \in S$ where $S$ is a subset of $k$-dimensional Euclidean space. The set of all *values* $\mathbf{s}$ with $\mu[\exp \mathbf{s}'\mathbf{Y}] < \infty$ is convex because exp is convex.

Exponential families are attractive for statistical analyses because they remain exponential under repeated sampling and under formation of posterior distributions. If $X$ is distributed as $E[\mu, \mathbf{Y}, \mathbf{s}]$, then the random sample $X_1, X_2, \ldots, X_n$ is distributed as $E[\mu^n, \sum \mathbf{Y}(X_i), \mathbf{s}]$. If $\mathbf{s}$ has prior probability $P$, and $X$ is distributed as $E[\mu, \mathbf{Y}, \mathbf{s}]$, the posterior probability of $\mathbf{s}$ given $X_1, X_2, \ldots, X_n$ is

$$P_\mathbf{x} = E[P, [s_1, s_2, \ldots, s_k, \lambda(\mathbf{s})], [\textstyle\sum Y_1(X_i), \ldots, \sum Y_k(X_i), n]]$$

where $\lambda(\mathbf{s}) = -\log P_\mathbf{s} \exp[\mathbf{s}'\mathbf{Y}(t)]$. Thus the posterior probabilities, for all data $X$, belong to the same exponential family! (This occurs for *all* prior probabilities $P$; there is NO special family of "conjugate" prior distributions!)

72

## 8.1. Examples of Exponential Families

(i) BERNOULLI $\qquad T = \{0, 1\}, \qquad \mu\{0\} = 1, \qquad \mu\{1\} = 1.$

$$P_p = p^t(1 - p)^{1-t} \quad \text{for } t = 0, 1, 0 \leq p \leq 1.$$

Let $s = \log p/(1 - p)$, $P_s = \exp[t \log p + (1 - t) \log(1 - p)]$

$$= \exp[ts + \lambda(s)], \quad \lambda(s) = -\log(1 + e^s).$$

For $n$ observations, this becomes the binomial.

(ii) POISSON $\qquad T = \{0, 1, \dots\}, \mu\{t\} = 1/t\,!.$

$$P_\lambda\{t\} = \lambda^t e^{-\lambda}/t! = \exp[t \log \lambda - \lambda]\mu\{t\}.$$

$$P_s\{t\} = \exp[ts - e^s] \quad \text{with } s = \log \lambda.$$

(iii) EXPONENTIAL $\qquad T = [0, \infty), \qquad \mu \text{ uniform on } T.$

$$f_\lambda(t) = e^{-t/\lambda}/\lambda.$$

$$f_s(t) = \exp[ts + \log(-s)], \qquad s = -1/\lambda.$$

(iv) NORMAL LOCATION $\qquad T = (-\infty, \infty), \qquad \mu \text{ unit normal.}$

$$f_\theta(t) = \exp[\theta t - \tfrac{1}{2}\theta^2], \qquad s = \theta.$$

(v) NORMAL SCALE $\qquad T = (-\infty, \infty), \qquad \mu \text{ uniform}$

$$f_\sigma(t) = \frac{1}{\sigma\sqrt{2\pi}} \exp[-\tfrac{1}{2}t^2/\sigma^2]$$

$$f_s(t) = \exp[st^2 + \tfrac{1}{2}\log[-s/\pi]],$$
$$s = -1/2\sigma^2.$$

[*Note.* The $\lambda(s)$ for the posterior density is the second expression in the exponential argument.]

[*Note.* Here parameters have been transformed to demonstrate reduction to exponential form; in computations, it is usually better to leave the parameters untransformed.]

## 8.2. Prior Distributions for the Exponential Family

If $P_s = E[\mu, Y, s]$, the *Jeffreys density* is $[\text{var } Y]^{1/2}$; only the binomial among the standard families has the Jeffreys density unitary.

Since $Y$ is of minimum variance among unbiased estimators of $P_s Y$, it is of interest to discover the prior distribution such that the posterior mean is $Y$. This prior distribution is necessarily not unitary.

**Theorem.** *Suppose* $P_s = E[\mu, Y, s]$, $s_1 < s < s_2$ *and that, for a given* $Y$, $\exp(Ys)/$
$\mu[\exp(Ys)] \to 0$ *as* $s \to s_1$ *or* $s \to s_2$. *The posterior mean, given* $Y$, *of* $P_s Y$ *is* $Y$
*if the prior probability of* $s$ *is uniform over* $(s_1, s_2)$.

PROOF.

$$P_s Y = \mu(Y \exp(Ys))/\mu(\exp(Ys))$$

$$= \frac{\partial}{\partial s} \log \mu[\exp(Ys)].$$

$$\int_{s_1}^{s_2} (P_s Y - Y)[\exp(Ys)/\mu(\exp Ys)]ds$$

$$= \int_{s_1}^{s_2} -\frac{\partial}{\partial s} \exp\{Ys - \log[\mu(\exp(Ys))]\}ds$$

$$= 0. \qquad\qquad \square$$

EXAMPLE. In the binomial case $t/n$ is the minimum variance unbiased estimate of $p$. If $s = \log(p/1 - p)$ is uniform over $-\infty, \infty$, then $t/n$ is the posterior mean of $p$, provided $0 < t < n$; in the case $t = 0, n$ the posterior mean is not defined (the limiting conditions of the theorem break down).

| Name | Family | Parameter | The Jeffreys Density (for $s$) |
|---|---|---|---|
| Binomial | $\binom{n}{t} p^t (1-p)^{n-t}$ | $s = \log \dfrac{p}{1-p}$ | $e^{(1/2)s}/(1 + e^s)$ |
| Poisson | $\lambda^t e^{-\lambda}/t!$ | $s = \log \lambda$ | $e^{(1/2)s}$ |
| Exponential | $e^{-t/\lambda}/\lambda$ | $s = -1/\lambda$ | $1/s$ |
| Normal Location | $\dfrac{1}{\sqrt{2\pi}} e^{\theta t - (1/2)\theta^2}$ | $s = \theta$ | $1$ |
| Normal Scale | $\dfrac{1}{\sigma\sqrt{2\pi}} \exp(-\tfrac{1}{2}t^2/\sigma^2)$ | $s = -1/2\sigma^2$ | $1/s$ |

## 8.3. Normal Location

Assume $X_1, \ldots, X_n, \theta$ are random variables such that $X_1, \ldots, X_n$ given $\theta$ are independent, $P_\theta^{X_i} = N(\theta, 1)$.

(i) *The posterior distribution.* If $\theta$ has prior probability $P$, the posterior probability $P_{X_1,\ldots,X_n}$ has density $\exp[-\tfrac{1}{2}n(\bar{X} - \theta)^2]/P \exp[-\tfrac{1}{2}n(\bar{X} - \theta)^2]$ with respect to $P$, where $\bar{X}$ denotes the mean of $X_1, \ldots, X_n$. The posterior probability is defined for $n$ large enough provided $P[\exp(-A\theta^2)] < \infty$ for some $A$. Say that such a $P$ is *docile*.

(ii) *Asymptotics.* If $\theta_0$ is in the support of a docile $P$, then $P_X$ eventually concentrates on $\theta_0$ as $P_{\theta_0}$. (For each $\varepsilon > 0$,

$$P_{\theta_0}\{P_X(|\theta - \theta_0| > \varepsilon) \nrightarrow 0\} = 0.)$$

If a docile $P$ has a density wrt Lebesgue measure that is continuous and positive at $\theta_0$, the posterior distribution is asymptotically normal $N(\bar{X}, 1/n)$

given $\theta_0$. (That is,

$$P_X(\theta \leqq \bar{X} + z/\sqrt{n}) \to \int_{-\infty}^{z} \frac{1}{\sqrt{2\pi}} \exp(-\tfrac{1}{2}u^2)du \quad [\text{in } P_{\theta_0}].)$$

If a docile $P$ has a density $p$ wrt Lebesgue measure that is continuously differentiable and positive at $\theta_0$, the posterior distribution is asymptotically $N(\bar{X} + [(\partial/\partial\theta_0) \log p]/n, 1/n)$ given $\theta_0$. (That is,

$$\sqrt{n}\left[ P_X\left[ \theta \leqq \bar{X} + \left[ \frac{\partial}{\partial\theta_0} \log p \right] \bigg/ n + z\sqrt{n} \right] - \int_{-\infty}^{z} \frac{1}{\sqrt{2\pi}} \exp(-\tfrac{1}{2}u^2)du \right] \to 0$$

$$[\text{in } P_{\theta_0}].)$$

Thus the principal effect asymptotically of a smooth prior density is a shift in location of the mean of the posterior distribution.

(iii) *The uniform prior.* The uniform prior is Lebesgue measure on the line. It is docile, the Jeffreys density, the only density invariant under location and sign changes, the only density for which the corresponding location estimate is unbiased.

The posterior distribution is $P_X^\theta = N(\bar{X}, 1/n)$. The posterior mean $\bar{X}$ is mean-square admissible, unbiased, of minimum variance among unbiased estimators.

The high density regions $\{\theta \,|\, n(\theta - \bar{X})^2 < k\}$ of posterior probability $\alpha$ are confidence regions of confidence size $\alpha$.

To test $\theta < 0$ against $\theta \geqq 0$, the Bayes decision accepts $\theta < 0$ if $P_X(\theta < 0) > k$, which is equivalent to $\bar{X} < c$, the uniformly most powerful test of $\theta < 0$ against $\theta \geqq 0$. If $\bar{X}_0$ is observed, it is customary to report the tail probability $P_{\theta=0}(\bar{X} > \bar{X}_0)$ in testing $\theta < 0$ against $\theta \geqq 0$; this is the same as $P_{X_0}(\theta < 0)$, the posterior probability of the null hypothesis.

To test $\theta = 0$ against $\theta \neq 0$, the Bayes test is of form accept $\theta = 0$ if $|\bar{X}| < c$, which is the most powerful unbiased test. If $\bar{X}_0$ is observed, it is customary to report the tail probability $P_{\theta=0}(|\bar{X}| > |\bar{X}_0|)$, which is the same as $P_X(|\bar{X}_0 - \theta| < |\bar{X}_0|)$, the posterior probability that the true mean is farther away from the observed mean than 0.

(iv) *Normal priors.* If $P = N(\theta_0, \sigma_0^2)$, then $P_X^\theta = N\{[(\theta_0/\sigma_0^2) + n\bar{X}]/((1/\sigma_0^2) + n), ((1/\sigma_0^2) + n)^{-1}\}$; the posterior mean is of the same form as the prior. (This is true for any prior; see 8.0.)

The formulae for means and variances may be remembered by the following scheme:

PRIOR: $\qquad\qquad \theta \sim N(\theta_0, \sigma_0^2)$
OBSERVATION: $\bar{X} \sim N(\theta, 1/n)$

Act as if $\theta_0$ is an observation on $\theta$; combine with the observation $\bar{X}$ inversely weighting by variances:

$$1/\text{var}_X(\theta) = 1/\text{var}_\theta(\bar{X}) + 1/\text{var}\,\theta$$
$$P_X(\theta)/\text{var}_X(\theta) = \bar{X}/\text{var}_\theta \bar{X} + \theta_0/\text{var}\,\theta.$$

It may happen that the prior $N(\theta_0, \sigma_0^2)$ and the observed $\bar{X}$ contradict each other. Note that $\bar{X} - \theta_0 \sim N(0, \sigma_0^2 + 1/n)$. Thus if $(\bar{X} - \theta_0)^2/(\sigma_0^2 + 1/n)$ is very large, we might decide to revise the observation $\bar{X}$, its distribution given $\theta$, or the prior for $\theta$. The contradiction will not arise if $\sigma_0^2$ is very large.

The posterior distribution $P_{\mathbf{X}}$ approaches the posterior distribution $N(\bar{X}, 1/n)$ corresponding to the uniform as $\sigma_0 \to \infty$, and so this family of priors is useful in showing admissibility of classical statistical procedures corresponding to the uniform.

(v) *Two stage normal priors.* Consider a family of normal priors $P_\lambda = N[\theta(\lambda), \sigma^2(\lambda)]$. Given $\lambda$, the posterior distribution $P_{\lambda, X}$ is normal with parameters given in (iv). Suppose that $\lambda$ itself has a prior distribution $Q$; then the prior distribution on $\theta$ is $Q(P_\lambda)$, a mixture of normal priors. The posterior distribution corresponding to $Q(P_\lambda)$ is also a mixture of normals $Q_{\bar{X}}(P_{\lambda, \bar{X}})$ where $Q_{\bar{X}}$ is the posterior distribution for $\lambda$ for the prior $Q$ and for the observation $\bar{X} \sim N[\theta(\lambda), \sigma^2(\lambda) + 1/n]$. [Given $\lambda$, $\theta$ is distributed as $N[\theta(\lambda), \sigma^2(\lambda)]$ and $\bar{X}$ is distributed as $N(\theta, 1/n)$; ignoring $\theta$, $\bar{X} \sim N(\theta(\lambda), \sigma^2(\lambda) + 1/n)$.] Thus values of $\lambda$ for which $[\bar{X} - \theta(\lambda)]^2/[\sigma^2(\lambda) + 1/n]$ is large will be downweighted, and contradiction between the prior mean $\theta(\lambda)$ and the observed $\bar{X}$ is prevented.

## 8.4. Binomial

The number of successes $t, 0 \le t \le n$ has probability $\binom{n}{t} p^t (1 - p)^{n-t}$. The prior $P$ is *docile* if $p^A(1 - p)^B$ is integrable for some $A, B$.

(i) *The posterior density* with respect to the docile prior $P$ is $p^t(1 - p)^{n-t}/P[p^t(1 - p)^{n-t}]$. The posterior density concentrates with probability 1 on $p_0$ if $p_0$ is true and $p_0$ lies in the support of $P$. If $P$ has a density that is continuous and positive at $p_0$, then $P_t$ is asymptotically normal $N[t/n, p_0(1 - p_0)/n]$.

(ii) *Beta priors.* If $P$ is $\text{Be}(\alpha, \beta)$, having density $p^{\alpha-1}(1 - p)^{\beta-1}/B(\alpha, \beta)$ wrt Lebesgue measure, then $P_t$ is $\text{Be}(\alpha + t, \beta + n - t)$. Note that if $P = \text{Be}(\alpha, \beta)$, then $Pp = \alpha/(\alpha + \beta)$, var $p = \alpha\beta/[(\alpha + \beta)^2(\alpha + \beta + 1)]$. The Jeffreys density is $\text{Be}(1/2, 1/2)$. The "unbiased" prior, $P = \text{Be}(0, 0)$ has posterior mean $P_t s = t/n$ for $0 < t < n$; the posterior distribution is not defined for $t = 0$, $t = n$. The admissibility of $t/n$ may be demonstrated by considering it as a limit of the Bayes posterior means $(t + \alpha)/(n + 2\alpha)$ corresponding to priors $\text{Be}(\alpha, \alpha)$ as $\alpha \to 0$.

(iii) *Confidence properties of beta priors.* The discreteness of $t$ makes it impossible to achieve unbiasedness of two sided tests $p = p_0$ against $p \ne p_0$, or to find set selection procedures that have the confidence property. Welch and Peers (1963) show that the Jeffreys density generates Bayes one sided intervals which are most nearly confidence intervals; but their proof is invalid for discrete observations.

For a particular prior, consider the one sided Bayes intervals $[0, p_t]$, such that $P_t\{p \leq p_t\} = \alpha, 0 \leq t \leq n$. The confidence properties of such intervals are determined by the function $P_p(p \leq p_t)$; this function is discontinuous in general at $(p_0, p_1, \ldots, p_n)$.

For the prior $\text{Be}(0, 1)$, $P_p(p \leq p_t) \leq \alpha$ all $p > 0$ and for the prior $\text{Be}(1, 0)$, $P_p(p \leq p_t) \geq \alpha$ all $p < 1$ (Thatcher, 1964). These might be viewed as "liberal" and "conservative" confidence regions for $p$. The inequalities follow from the identity:

$$\sum_{t=t_0}^{n} \binom{n}{t} p^t (1-p)^{n-t} = \int_0^p p^{t_0-1}(1-p)^{n-t_0} dp / B(t_0, n - t_0 + 1)$$

which relates binomial and beta (confidence and Bayes) probabilities.

For $\text{Be}(1, 0)$, $P_p(p \leq p_t) \downarrow \alpha$ as $p \uparrow p_k$, $0 < k \leq n$.

For $\text{Be}(0, 1)$, $P_p(p \leq p_t) \uparrow \alpha$ as $p \downarrow p_k$, $0 \leq k < n$.

Let $p_{t,\lambda}$ be such that $P_t(p \leq p_{t,\lambda}) = \alpha$ for the prior $\text{Be}[\lambda, 1 - \lambda], 0 \leq \lambda \leq 1$. Then $p_{t,\lambda}$ is increasing in $\lambda$. [The prior is equivalent to observing $\lambda$ successes and $1 - \lambda$ failures; the larger $\lambda$, the more the posterior for a particular $t$ is shifted to the right.] Thus for each prior $\text{Be}[\lambda, 1 - \lambda], 0 \leq \lambda \leq 1$, $\lim_{p \uparrow p_{t,\lambda}} P_p(p \leq p_{t,\lambda}) \geq \alpha$, $\lim_{p \downarrow p_{t,\lambda}} P_p(p \leq p_{t,\lambda}) \leq \alpha$. The confidence values cross the correct probability $\alpha$ at each of the points of discontinuity $p_{t,\lambda}$.

In Figure 1, $n = 10$, $\alpha = 0.9$ and the upper and lower bounding confidence curves for $\text{Be}(0, 1)$ and $\text{Be}(1, 0)$ are given, together with the intermediate curve for $\text{Be}(1/2, 1/2)$, the Jeffreys density. Note that $P_p^t(p \leq p_t) \to 1$ as $p \to 0$, $P_p^t(p \leq p_t) \to 0$ as $p \to 1$ for the Jeffreys density, so that it can never give confidence values uniformly near $\alpha$. It does give confidence values which are closer on the average to the correct $\alpha$ than the bounding priors $\text{Be}(0, 1)$ and $\text{Be}(1, 0)$.

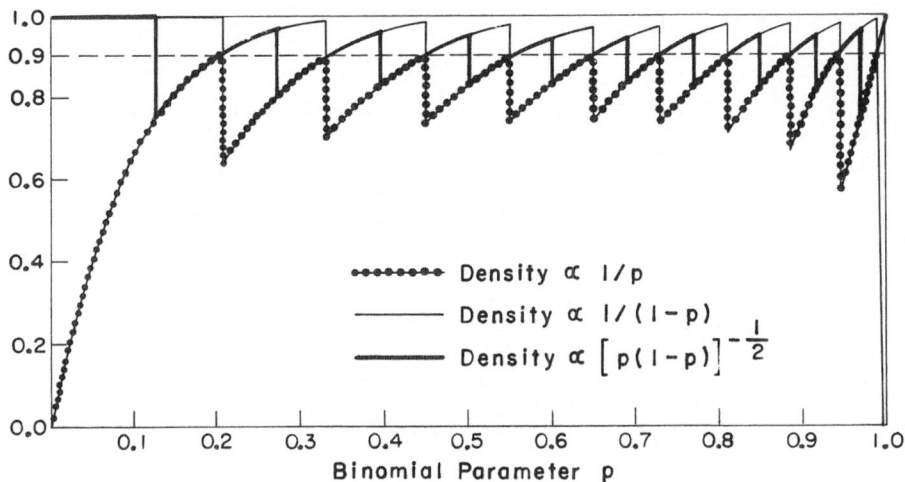

Figure 1.

An arbitrary interval selection procedure specifies an interval $\{p \leq p_t\}$ for each of $t = 0, 1, \ldots, n$. Its confidence properties are given by the function $P_p\{p \leq p_t\}$, which is discontinuous at each of the points $p_0 < p_1 < \cdots < p_n$. The overall error of the procedure might be assessed by $\sup\limits_{\varepsilon < p < 1 - \varepsilon} |P_p(p \leq p_t) - \alpha|$; it is necessary to bound $p$ away from 0 and 1, because 0 lies in every interval and 1 usually lies in no intervals, so that $P_0(p \leq p_t) = 1$, $P_1(p \leq p_t) = 0$. The maximum error is achieved at the points of discontinuity; it will be minimized by ensuring that $\frac{1}{2}(\lim\limits_{p \uparrow p_s} P_p(p \leq p_t) + \lim\limits_{p \downarrow p_s} P_p(p \leq p_t)) = \alpha$ at each point of discontinuity $p_s$. In this case, the asymptotic error at $p_s$ is $(1/2\sqrt{2\pi})\exp(-\frac{1}{2}Z_\alpha^2).(1/\sqrt{p_s(1 - p_s)}).(1/\sqrt{n})$ where $Z_\alpha$ is such that $P(Z \leq Z_\alpha) = \alpha$ for a normally distributed $Z$. (This result is obtained by equating binomial and beta tail areas and then using Edgeworth expansions for the beta distribution, involving the first three moments.) For a Bayes procedure, the asymptotic error is $(1/\sqrt{2\pi}) \exp(-\frac{1}{2}Z_\alpha^2)(1/\sqrt{p_s(1 - p_s)}).(1/\sqrt{n}) \sup(|\varDelta|, |\varDelta - 1|)$, where the prior density $h$ satisfies $\varDelta = [(\partial/\partial p) \log(h(p))p(1 - p)]_{p = p_s}$. This error is minimized for all $p_s$, precisely when $h = j$, the Jeffreys density, and in this case the interval selection procedure is close as possible to being a confidence procedure. (The error $\sup\limits_{\varepsilon < p < 1 - \varepsilon} |P_p(p \leq p_t) - \alpha|$ is $O(n^{-1/2})$ for every prior, but is smallest for the Jeffreys prior. If $p_0, p_1, \ldots, p_n$ are the upper bounds of intervals taken to ensure $\frac{1}{2}[\lim\limits_{p \uparrow p_s} P_p(p \leq p_t) + \lim\limits_{p \downarrow p_s} P_p(p \leq p_t)] = \alpha$ for each $p_s$, and if $p'_s$ denote the Bayes upper bounds, then $p'_s = p_s + O(1/n)$ for any Bayes procedure, and $p'_s = p_s + o(1/n)$ for the Jeffreys density.)

In the table below, the intervals $\{p \leq p_t\}$ are specified corresponding to the three priors $Be(0, 1)$, $Be(1, 0)$, $Be(\frac{1}{2}, \frac{1}{2})$, and also for a confidence procedure minimizing maximum error.

EXAMPLE. For $n = 10$, $\alpha = .90$, the intervals for various methods:

|          | $Be(0, 1)$ | $Be(1, 0)$ | $Be(\frac{1}{2}, \frac{1}{2})$ | Confidence |
|----------|-----------|-----------|-------------------------------|------------|
| $p_0$    | .0        | .2057     | .1236                         | .1487      |
| $p_1$    | .2057     | .3368     | .2744                         | .2981      |
| $p_2$    | .3368     | .4496     | .3948                         | .4063      |
| $p_3$    | .4496     | .5517     | .5018                         | .5118      |
| $p_4$    | .5517     | .6458     | .5997                         | .6090      |
| $p_5$    | .6458     | .7327     | .6901                         | .6990      |
| $p_6$    | .7327     | .8124     | .7735                         | .7823      |
| $p_7$    | .8124     | .8842     | .8494                         | .8584      |
| $p_8$    | .8842     | .9455     | .9164                         | .9257      |
| $p_9$    | .9955     | .9895     | .9704                         | .9799      |
| $p_{10}$ | .9895     | 1.0000    | .9998                         | 1.0000     |

## 8.5. Poisson

The number of occurrences $t, 0 \leqq t < \infty$, has probability $P_\lambda\{t\} = \lambda^t e^{-\lambda}/t!$. The prior $P$ is *docile* if $\lambda^K e^{-\lambda}$ is integrable some $K$.

(i) *The posterior density.* with respect to the prior probability $P$ is $\lambda^t e^{-\lambda}/P(\lambda^t e^{-\lambda})$. The posterior density concentrates on $\lambda_0$ with probability 1, if $\lambda_0$ is true and lies in the support of $P$. If $P$ has a continuous positive density at $\lambda_0$, then $P_t$ is asymptotically normal $N(\lambda_0, \lambda_0/n)$.

(ii) *Gamma priors.* The prior $G(m, a)$ has density $a^m \lambda^{m-1} e^{-a\lambda}/\Gamma(m)$, the gamma density. The posterior given $t$ is $G(m + t, a + 1)$. The Jeffreys density is $G(\frac{1}{2}, 0)$, not unitary. The "unbiased" prior is $G(0, 0)$, which has posterior mean $t$; the posterior distribution is not defined for $t = 0$.

(iii) *Confidence properties of gamma priors.* Similar considerations to those for the binomial apply. Exact confidence intervals are not possible because of the discreteness of the Poisson. For a particular prior $P$, let $0 \leq \lambda \leq \lambda_t$ be the $\alpha$-probability interval, $P_t(0 \leq \lambda \leq \lambda_t) = \alpha$. The confidence function $P(\lambda \leq \lambda_t)$ will be discontinuous at $\lambda_0, \lambda_1, \ldots$.

For the prior $G(0, 0)$, $P_\lambda(\lambda \leq \lambda_t) \leq \alpha$ all $t$, and for the prior $G(1, 0)$, $P_\lambda(\lambda \leq \lambda_t) \geq \alpha$ all $t$. These results follow from the equivalence between Poisson and gamma tails:

$$\sum_{t=t_0}^{\infty} \frac{\lambda^t}{t!} e^{-\lambda} = \int_0^{\infty} \frac{x^{t_0-1}}{(t_0 - 1)!} e^{-x} dx.$$

As in the binomial case, Jeffreys' density gives intervals which are closest to confidence intervals in that

$$\tfrac{1}{2}(\lim_{\lambda \uparrow \lambda_s} P_\lambda(\lambda \leq \lambda_t) + \lim_{\lambda \downarrow \lambda_s} P_\lambda(\lambda \leq \lambda_t))$$

is closest to $\alpha$ at every $\lambda_s$.

## 8.6. Normal Location and Scale

Suppose $X_1, X_2, \ldots, X_n$ are from $N(\mu, \sigma^2)$, where $\mu$ and $\sigma^2$ are unknown; then $X_1, \ldots, X_n$ is $E\left[v^n \left(\dfrac{\sum X_i}{\sum X_i^2}\right), \left(\dfrac{\mu/\sigma^2}{-1/2\sigma^2}\right)\right]$ where $v^n$ is Lebesgue measure on $R^n$. The prior $P$ is *docile* if $\exp[-(A + B\mu^2)/\sigma^2]$ is integrable for some $A, B > 0$.

(i) *General priors.* For the prior $P$, the posterior $P_t$ has density $\sigma^{-n} \exp[-\frac{1}{2}\sum X_i^2/\sigma^2 + \sum X_i \mu/\sigma^2 - \frac{1}{2}n\mu^2/\sigma^2]k(X_1, \ldots, X_n)$ with respect to $P$. If $(\mu_0, \sigma_0^2)$ lies in the support of $P$, and $\mu_0, \sigma_0^2$ is the true value, then the posterior distribution concentrates on $(\mu_0, \sigma_0^2)$ with probability 1. If $P$ has a positive continuous density at $(\mu_0, \sigma_0^2)$, then the posterior density is asymptotically normal.

(ii) *Invariance generated priors.* A prior with density [with respect to Lebesgue measure on $\mu, \sigma^2$] $\sigma^{-A} \exp(-\frac{1}{2}B/\sigma^2 + C\mu/\sigma^2 - \frac{1}{2}D\mu^2/\sigma^2 + K)$ is called an invariance generated prior $IG(A, B, C, D)$. After the observations $X_1, X_2, \ldots, X_n$, the posterior is $IG(A + n, B + \sum X_i^2, G + \sum X_i, D + n)$. Priors of the form $IG(A, 0, 0, 0)$ are improper, invariant priors under the transformations $X_i \to a + bX_i, \mu \to a + b\mu, \sigma \to |b|\sigma$; by considering posterior distributions obtained from invariant priors, for various types of data, we obtain the distributions $IG(A, B, C, D)$ where $B \geq 0, D$ is an integer, $C^2 \leq BD$.

The Jeffreys density is $IG(2, 0, 0, 0)$.

The density $IG(5, 0, 0, 0)$, for parameters $(1/n)P_{\mu,\sigma^2}(\sum X_i) = \mu$ and $(1/n)P_{\mu,\sigma^2}(\sum X_i^2) = \sigma^2 + \mu^2$, has posterior means $(1/n)\sum X_i$ and $(1/n)\sum X_i^2$; this corresponds to $(\mu/\sigma^2, -1/2\sigma^2)$ being uniform in the plane, see 8.2.

(iii) *Marginal distributions of $\mu$ and $\sigma^2$.* The marginal density of $\mu$ corresponding to $IG(A, B, C, D)$ is $K_1(\frac{1}{2}B - C\mu + \frac{1}{2}D\mu^2)^{-(A-1)/2}$, which is a student distribution with $A - 2$ degrees of freedom. (The conditional density of $\mu$ given $\sigma^2$ is normal.)

The marginal density of $1/\sigma^2 = u$ corresponding to $IG(A, B, C, D)$ is $K_2 u^{(A-2)/2-1} \exp[-\frac{1}{2}(B - D^2/C)u]$, which is a gamma distribution with $A - 2$ degrees of freedom.

(iv) *The "confidence" prior* $IG(1, 0, 0, 0)$. For this prior, the posterior is $IG(n + 1, \sum X_i^2, \sum X_i, n)$, and the marginal densities of $\mu$ and $\sigma$ are:

$$\sqrt{n}(\bar{X} - \mu)/s \sim T_{n-1}$$

$$(n - 1)s^2/\sigma^2 \sim \chi_{n-1}^2$$

where $\bar{X} = 1/n\sum X_i$, $s^2 = \sum(X_i - \bar{X})^2/(n - 1)$, $T_{n-1}$ denotes a student distribution on $(n - 1)$ degrees of freedom, $\chi_{n-1}^2$ denotes a chi-square distribution on $(n - 1)$ degrees of freedom.

Since the same distributions hold when $\mu$ and $\sigma^2$ are fixed, and $\bar{X}$ and $s$ are random, Bayes intervals and regions have a confidence interpretation. For example, the high density region of $\mu$, $\{\mu|\sqrt{n}|\bar{X} - \mu| \leq sT_{n-1,\alpha}\}$ has posterior probability $\alpha$ given $X_1, \ldots, X_n$, but also probability $\alpha$ given $\mu, \sigma^2$. (Here $P(|T_{n-1}| \leq T_{n-1,\alpha}) = \alpha$.) Or, the one-sided Bayes interval for $\sigma^2$, $\{\sigma^2|(n - 1)s^2/\sigma^2 \leq \chi_{n-1,\alpha}^2\}$ has posterior probability $\alpha$ of containing $\sigma^2$, but also probability $\alpha$ given $\sigma^2$. (Here $P(\chi_{n-1}^2 \leq \chi_{n-1,\alpha}^2) = \alpha$.)

(It should be noted that the high posterior density region for $\mu$ and $\sigma^2$ is not a confidence region for this prior; the Jeffreys prior $IA(2, 0, 0, 0)$ gives a high posterior density region

$$(s/\sigma)^n \exp[-\frac{1}{2}(n - 1)s^2/\sigma^2 - \frac{1}{2}n(\bar{X} - \mu)^2/\sigma^2] \geq c$$

which is also a confidence region.)

The Bayes test for $\mu = \mu_0$ against $\mu \neq \mu_0$ is: accept $\mu = \mu_0$ if $|\bar{X} - \mu_0| < cs$. The tail area $P[|T_{n-1}| \geq \sqrt{n}|\bar{X} - \mu_0|/s]$ is the Bayes posterior probability

$P_X[|\mu - \bar{X}| \geq |\mu_0 - \bar{X}|]$, the probability that $\mu$ is further from the observed $\bar{X}$ than $\mu_0$.

(v) *Unbetworthiness of confidence interval for* $\mu$. The interval $\sqrt{n}|\mu - \bar{X}| \leq T_{n-1,\alpha}s$, where $P(|T_{n-1}| < T_{n-1,\alpha}) = \alpha$ is not betworthy. If $s < 1$, bet $1 - \alpha$ to receive 1 if $\sqrt{n}|\mu - \bar{X}| > T_{n-1,\alpha}s$. If $s \geq 1$, bet $\alpha$ to receive 1 if $\sqrt{n}|\mu - \bar{X}| \leq T_{n-1,\alpha}s$. The strategy is to bet that $\mu$ does not lie in the interval when $s$ is small, and to bet that $\mu$ does lie in the interval when $s$ is large. Since $P[\sqrt{n}|\bar{X} - \mu| \leq T_{n-1,\alpha}s|s, \mu, \sigma]$ increases strictly with $s$, and averages $\alpha$ over all $s$,

$$P[\sqrt{n}|\bar{X} - \mu| \leq T_{n-1,\alpha}s|s < 1] \leq \alpha < P[\sqrt{n}|\bar{X} - \mu| \leq T_{n-1,\alpha}s|s > 1].$$

The above bet will always have positive expectation, no matter what the value of $\mu, \sigma$. However, as $\sigma^2 \to 0$ or $\infty$, the net gain from the bet will be arbitrarily close to zero.

More generally, bet $\alpha k(s)$ to receive $k(s)$ if $\sqrt{n}|\mu - \bar{X}| \leq T_{n-1,\alpha}s$, where $k(s)$ may be negative. Whenever the function $k(s)$ is strictly increasing in $s$, the net gain from the bet is $P_{\mu,\sigma}[k(s)[\{\sqrt{n}[\mu - \bar{X}] \leq T_{n-1,\alpha}s\} - \alpha]] > 0$. For $k(s) = s^\lambda$, the net gain is $\sigma^\lambda K(n, \alpha, \lambda)$ where $K(n, \alpha, \lambda)$ has the same sign as $\lambda$. Thus the bet $s^2/K(n, \alpha, 2) + s^{-2}/K(n, \alpha, -2)$ has gain $\sigma^2 + \sigma^{-2} \geq 2$. It is thus possible to devise bets whose net gain is greater than 2 for all $\sigma^2$. See Brown (1967).

(vi) *The Behrens–Fisher problem.* Suppose that $X_1, \ldots, X_n$ are a sample from $N(\mu_1, \sigma_1^2)$, and $Y_1, \ldots, Y_m$ are a sample from $N(\mu_2, \sigma_2^2)$. Taking the "confidence" prior density (with respect to Lebesgue measure $v$ on $\mu_1, \mu_2, \sigma_1, \sigma_2) \sigma_1^{-1}\sigma_2^{-1}$, and letting

$$\bar{X} = \frac{1}{n}\sum X_i, s_X^2 = \sum(X_i - \bar{X})^2/(n - 1),$$

$$\bar{Y} = \frac{1}{m}\sum Y_i, s_Y^2 = \sum(Y_i - \bar{Y})^2/(m - 1),$$

the posterior distributions of $\mu_1$ and $\mu_2$ are independently

$$\mu_1 \sim \bar{X} + s_X T_{n-1}/\sqrt{n},$$
$$\mu_2 \sim \bar{Y} + s_Y T_{m-1}/\sqrt{m}.$$

Then $\mu_1 - \mu_2$ is the convolution of two student distributions.

In order to test $\mu_1 = \mu_2$ against $\mu_1 \neq \mu_2$, Behrens and Fisher propose the test which rejects $\mu_1 = \mu_2$ if

$$P_{X,Y}[|\mu_1 - \mu_2 - (\bar{X} - \bar{Y})| > |\bar{X} - \bar{Y}|]$$ is sufficiently small.

The Bayes test would reject $\mu_1 = \mu_2$ if the posterior density of $\mu_1 - \mu_2$ at 0 is sufficiently small—that is, if

$$\int \left[ \sum_{i=1}^{m} (X_i - v)^2 \right]^{-(m+1)/2} \left[ \sum_{i=1}^{n} (Y_i + v)^2 \right]^{-(n+1)/2} dv$$

is small enough.

Both tests have probability of rejection, given $\mu_1 = \mu_2$, that depends on $\sigma_1/\sigma_2$.

## 8.7. Problems

E1. Suppose $X_1, \ldots, X_n$ is a random sample from $N(\theta, 1)$, and that the prior distribution $P$ has $P\{-1\} = P\{1\} = \frac{1}{2}$. If $\theta = 0$, what is the asymptotic behavior of the posterior distribution?

E2. If $X$ is an observation from $N(\theta, 1)$, show that for every unitary prior $P$, $PP_\theta[\theta - P_X(\theta)]^2 < 1$.

P1. If $X$ is an observation from $N(\theta, 1)$, show that $aX$ is an admissible estimate of $\theta$ using the loss function $L(d, \theta) = (d - \theta)^2$, for $0 \leq a \leq 1$.

P2. If $X$ is Poisson with parameter $\lambda$, and the prior on $\lambda$ is gamma, $G(m, a)$, find the Bayes estimate of $\lambda$ with loss $L(d, \lambda) = (d/\lambda - \lambda/d)^2$.

Q1. In a test of 10 questions, a child gets $t$ questions correct where $t$ is binomial with parameter $p$. Over many children, the parameter $p$ has prior distribution $P$. The observed number of successes over many children is:

| NUMBER CORRECT | 0 | 1 | 2 | 3 | 4 | 5 | 6 | 7 | 8 | 9 | 10 |
|---|---|---|---|---|---|---|---|---|---|---|---|
| NUMBER OF CHILDREN | 66 | 240 | 540 | 960 | 2450 | 3016 | 2520 | 2520 | 2970 | 2640 | 1716 |

Estimate $P$.

P3. If an observation $t$ has probability $P_s = E[\mu, Y, s]$, show that Bayes tests of $s \leq s_0$ against $s > s_0$ are of form: decide $s \leq s_0$ if $Y \leq Y_0$.

P4. If $X_1, \ldots, X_n$ are observations from $N(\theta, 1)$, and for the uniform prior on $\theta$, find the conditional distribution of $X_{k+1}, \ldots, X_n$ given $X_1, X_2, \ldots, X_k$.

E3. For a normal sample $X_1, \ldots, X_n$ from $N(\mu, \sigma^2)$, with the prior IG(1, 0, 0, 0), find the posterior mode of $\mu$ and $\sigma^2$, and the posterior means of $\mu$ and $\sigma^2$, based on the posterior density of $\mu$ and $\sigma^2$.

P5. For a normal sample $X_1, \ldots, X_n$ from $N(\mu, \sigma^2)$, with prior IG(1, 0, 0, 0) find the Bayes estimator of $\sigma^2$ using loss function $L(d, \sigma^2) = (d - \sigma^2)^2$, and compare its risk function with those of maximum likelihood and unbiased estimates of $\sigma^2$.

## 8.8. References

Brown, L. (1967), The conditional level of the t-test, *Annals of Mathematical Statistics* **38**, 1068–1071.

Thatcher, A. R. (1964), Relationships between Bayesian and confidence limits for prediction, *J. R. Statist. Soc.* B **26**, 176–210.

Welch, B. L. and Peers, H. W. (1963), On formulae for confidence points based on intervals of weighted likelihoods, *J. Roy. Statist. Soc.* B **25**, 318–329.

# CHAPTER 9

# Many Normal Means

## 9.0. Introduction

Given $\mathbf{X}$, suppose $P_{\mathbf{X}}^{Y_i} = N(X_i, 1)$, $i = 1, 2, \ldots, n$, and the $Y_i$ are independent. The straight estimate $Y_i$ of $X_i$ is least squares, maximum likelihood, of minimum variance among unbiased estimators, posterior means with respect to the Jeffreys density (the $X_1, \ldots, X_n$ are uniform) but for all these virtues inadmissible with loss function $\sum_{i=1}^{n} (d_i - X_i)^2$ for $n > 2$, Stein (1956).

## 9.1. Baranchik's Theorem

**Lemma.** If $Y \sim N(X, 1)$, $X \geq 0$, and if $f$ is integrable

$$P[f(Y^2)] = \sum_{k=0}^{\infty} p_k P[f(\chi^2_{2k+1})]$$

$$P[Yf(Y^2)] = X \sum p_k P[f(\chi^2_{2k+3})]$$

where $\chi^2_{2k+1}$ denotes a variable with the chi-square distribution, and

$$p_k = \exp(-\tfrac{1}{2}X^2)(\tfrac{1}{2}X^2)^k / k!$$

are Poisson probabilities with expectation $\tfrac{1}{2}X^2$.

PROOF. The first result, that a non-central chi-square is a mixture of central chi-squares with Poisson mixing probabilities, should have a nice probabilistic proof, but I don't know one.

84

(i) $P[f(Y^2)] = \int_{-\infty}^{\infty} f(Y^2) \exp[-\frac{1}{2}(X-Y)^2] dY/\sqrt{2\pi}$

$\qquad = \int f(Y^2) \exp(-\frac{1}{2}Y^2) \exp[XY] dY \exp[-\frac{1}{2}X^2]/\sqrt{2\pi}$

$\qquad = \int_{-\infty}^{\infty} f(Y^2) \sum_{k=0}^{\infty} \frac{X^k Y^k}{k!} \exp[-\frac{1}{2}Y^2] dY \exp[-\frac{1}{2}X^2]/\sqrt{2\pi}.$

{Note $\chi_{2k+1}^2 = Y^2 = u$ has density: $u^{k-1/2} \exp(-\frac{1}{2}u)(\frac{1}{2})^{k+1/2}/\Gamma(k+\frac{1}{2})$.}

$P[f(Y^2)] = 2\int_0^{\infty} f(Y^2) \sum_{k=0}^{\infty} \frac{X^{2k} Y^{2k}}{2k!} \exp[-\frac{1}{2}Y^2] dY \exp[-\frac{1}{2}X^2]/\sqrt{2\pi}$

$\qquad = 2\int_0^{\infty} f(Y^2) \sum \frac{(\frac{1}{2}X^2)^k (\frac{1}{2}Y^2)^k}{k!(k-\frac{1}{2})(k-\frac{3}{2})\cdots\frac{1}{2}}$

$\qquad\quad \cdot \exp[-\frac{1}{2}Y^2] dY \exp[-\frac{1}{2}X^2]/\sqrt{2\pi}$

$\qquad = \sum \int_0^{\infty} f(Y^2) p_k \frac{(\frac{1}{2}Y^2)^{k-1/2}}{\Gamma(k+\frac{1}{2})} \exp[-\frac{1}{2}Y^2] \frac{dY^2}{2} [\text{since } \Gamma(\frac{1}{2}) = \sqrt{\pi}]$

$\qquad = \sum p_k P[f(\chi_{2k+1}^2)].$

(ii) $P[Yf(Y^2)] = \int_{-\infty}^{\infty} f(Y^2) \sum \frac{X^k Y^{k+1}}{k!} \exp[-\frac{1}{2}Y^2] dY \exp[-\frac{1}{2}X^2]/\sqrt{2\pi}$

$\qquad = 2\int_0^{\infty} f(Y^2) \sum_0^{\infty} \frac{X^{2k+1} Y^{2k+2}}{(2k+1)!}$

$\qquad\quad \cdot \exp[-\frac{1}{2}Y^2] dY \exp[-\frac{1}{2}X^2]/\sqrt{2\pi}$

$\qquad = X \sum_{k=0}^{\infty} p_k P[f(\chi_{2k+3}^2)]$ after some algebra. $\qquad\square$

**Theorem** (Baranchik (1970)). *Let $Y_i$ be independent $N(X_i, 1)$, $i = 1, \ldots, n$. Let $S = \sum Y_i^2$, and let $f$ be a non-decreasing non-negative function with $f < 2(n-2)$. Then $P[\sum\{Y_i[1 - f(S)/S] - X_i\}^2] < P[\sum(Y_i - X_i)^2]$ for every $X_1, X_2, \ldots, X_n$, if $n > 2$.*

PROOF. Since $S = \sum_{i=1}^{n} Y_i^2$ is invariant under rotations of the $Y_i$, $\sum(Y_i(1 - f(S)/S) - X_i)^2$ has the same distribution if $\mathbf{Y}$ and $\mathbf{X}$ undergo the same rotation. It is sufficient therefore to consider $X_1 \geq 0$, $X_i = 0$ for every $i \geq 2$.

Let $g(S) = 1 - f(S)/S$

$$P\sum(Y_i g(S) - X_i)^2 = P[\sum Y_i^2 g^2(S) - 2X_1 Y_1 g(S) + X_1^2].$$

Then $S = Y_1^2 + Z$ where $Z \sim \chi_{n-1}^2$ independent of $Y_1$.

$$P[Sg^2(S)] = P^Z P_Z(Y_1^2 + Z)g^2(Y_1^2 + Z)$$

$$= P^Z \sum p_k P_z[(\chi_{2k+1}^2 + Z)g^2(\chi_{2k+1}^2 + Z)] \text{ from the lemma}$$

$$= \sum p_k P[(\chi_{2k+n}^2)g^2(\chi_{2k+n}^2)], \; p_k = \exp(-\tfrac{1}{2}X_1^2)(\tfrac{1}{2}X_1^2)^k/k!$$

$$P[Y_1 g(S)] = X_1 \sum p_k P[g(\chi_{2k+n+2}^2)] \text{ from the lemma}$$

$$P\sum[Y_i g(S) - X_i]^2 = \sum p_k P[\chi_{2k+n}^2 g(\chi_{2k+n}^2) - 2X_1^2 g(\chi_{2k+n+2}^2) + X_1^2]$$

$$= \sum_{}^{\infty} p_k P[\chi_{2k+n}^2 g^2(\chi_{2k+n}^2) - 4kg(\chi_{2k+n}^2) + 2k]$$

$$P[\sum(Y_i g(S) - X_i)]^2 - P\sum(Y_i - X_i)^2$$

$$= \sum p_k P[\chi_{2k+n}^2 g^2(\chi_{2k+n}^2) - 4kg(\chi_{2k+n}^2) + 2k - n].$$

This expression is to be shown $< 0$.

Set $g(S) = 1 - f(S)/S$ and note that

$$P[f(\chi_{2k+n}^2)/\chi_{2k+n}^2] \leqq Pf(\chi_{2k+n}^2)P\frac{1}{\chi_{2k+n}^2}$$

because $f$ is non-decreasing, so $f(\chi^2)$ and $1/\chi^2$ are negatively correlated.

$$P[\chi_{2k+n}^2 g^2(\chi_{2k+n}^2) - 4kg(\chi_{2k+n}^2) + 2k - n]$$

$$= Pf(\chi_{2k+n}^2)[-2 + f/\chi_{2k+n}^2 + 4k/\chi_{2k+n}^2]$$

$$< Pf(\chi_{2k+n}^2)\left[-2 + \frac{2(n-2) + 4k}{n + 2k - 2}\right] = 0.$$

since $f < 2(n - 2)$ and $P[1/\chi_{2k+n}^2] = 1/(n + 2k - 2)$. Thus $Yg(S)$ beats $Y$.  $\square$

*Note.* $Yg(S)$ shrinks the estimate towards 0, but the same result holds if it is shrunk towards any other point $Z$ by $Z + (Y - Z)g(S_z)$ where $S_z = \sum(Y_i - Z_i)^2$. Or shrunk towards $\bar{Y}1, f < 2(n - 3)$.

## 9.2. Bayes Estimates Beating the Straight Estimate

**Theorem.** *Suppose* $Y_i \sim N(X_i, \sigma^2)$, $i = 1, 2, \ldots, n$, $\sigma^2$ *known. Let the prior distribution for* $X_i$ *be* $X_i \sim N(0, \sigma_0^2)$ *independently given* $\sigma_0^2$, *where* $\sigma_0^2$ *has a density* $g$ *such that* $\log g$ *is concave in* $\log(\sigma^2 + \sigma_0^2)$ *and* $(\sigma^2 + \sigma_0^2)^{1-(1/2)\alpha}g$ *is increasing for some* $\alpha$. *The posterior mean of* $X_i$ *given* $Y_i$ *has smaller mean square error as an estimate of* $X_i$ *than* $Y_i$ *for every choice of* $X_1, \ldots, X_n$, *whenever* $n \geqq 4 - \alpha$.

PROOF. First fixing $\sigma_0^2$,

$$P[X_i | Y_i, \sigma_0^2] = Y_i \Big/ \left(1 + \frac{\sigma^2}{\sigma_0^2}\right).$$

$Y_i \sim N(0, \sigma^2 + \sigma_0^2)$ independently.

The posterior density of $\sigma_0^2$ given $Y_1, \ldots, Y_n$ is

$$g[\sigma_0^2|\mathbf{Y}] \propto [\sigma^2 + \sigma_0^2]^{-n/2} \exp\left[ -\tfrac{1}{2}\sum Y_i^2/(\sigma_0 + \sigma_0^2)\right]g(\sigma^2 + \sigma_0^2).$$

Letting $S = \sum Y_i^2$, $V = S/(\sigma^2 + \sigma_0^2)$,

$$P[V|\mathbf{Y}] = \int V^{n/2-1}\exp(-\tfrac{1}{2}V)g(S/V)dV/\int V^{n/2-2}\exp(-\tfrac{1}{2}V)g\left(\frac{S}{V}\right)dV$$

$$= P[\chi_{n-2}^2 g(S/\chi_{n-2}^2)]/P[g(S/\chi_{n-2}^2)].$$

Now
$$P[X_i|\mathbf{Y}] = Y_i P\left[\left(1 + \frac{\sigma^2}{\sigma_0^2}\right)^{-1}\Bigg|\mathbf{Y}\right]$$

$$= Y_i[1 - P(V|\mathbf{Y})\sigma^2/S].$$

From Theorem 9.1, the estimate $P[X_i|\mathbf{Y}]$ will beat $Y_i$ if $P(V|S) = P(V|\mathbf{Y})$ is a non-negative, non-decreasing function of $S\sigma^2$ such that $P(V|S) < 2(n-2)$. It is obviously non-negative.

Let $k(V) = V^{n/2-2}\exp(-\tfrac{1}{2}V)$. For $S > S'$

$$P[V|S]/P[V|S'] = \frac{\int V k(V)g(S/V)k(U)g(S'/U)dVdU}{\int U k(U)g(S'/U)k(V)g(S/V)dVdU} \geq 1$$

if $\int(V - U)k(V)k(U)\left[g(S/V)g(S'/U) - g(S/U)g(S'/V)\right]dUdV \geq 0$. Since $\log g$ is concave in $\log(\sigma^2 + \sigma_0^2)$, $g(S/V)g(S'/U) - g(S/U)g(S'/V) \geq 0$ for $S \geq S'$, $V \geq U$. Thus $P[V|S]$ is increasing in $S$.

Since $(\sigma^2 + \sigma_0^2)^{1-(1/2)\alpha} g = h$ is increasing,

$$P[V|S] = \int V^{(n-\alpha)/2}\exp(-\tfrac{1}{2}V)h\left(\frac{S}{V}\right)dV/\int V^{(n-\alpha)/2}\exp(-\tfrac{1}{2}V)h\left(\frac{S}{V}\right)dV$$

$$= P[\chi_{n-\alpha}^2 h(S/\chi_{n-\alpha}^2)]/P[h(S/\chi_{n-\alpha}^2)]$$

$$\leq P(\chi_{n-\alpha}^2) = n - \alpha \leq 2(n-2) \text{ if } n \geq 4 - \alpha.$$

Thus $P[V|S]$ satisfies the conditions of Theorem 9.1 and the theorem is proved.                                                                      □

*Note.* Priors of the above type will be unitary only if $\alpha < 0$, so that for a unitary Bayes estimate $n \geq 5$ is required to beat the straight estimate.

For $\alpha < 2$, the loss $\sum(X_i - P[X_i|\mathbf{Y}])^2$ is integrable, so the posterior mean is Bayes and hence admissible; a Bayes estimate may thus be obtained for $n \geq 3$. Strawderman (1971) considers the densities $g(\sigma_0^2) \propto (\sigma_0^2 + \sigma^2)^{(1/2)\alpha-1}$; then $P[V|S] = P[\chi_{n-\alpha}^2|\chi_{n-\alpha}^2 < S/\sigma^2]$. The particular choice $\alpha = 0$ is suggested by Jeffreys (1961).

James and Stein (1961) showed that estimates $Y_i(1 - ((n-2)\sigma^2)/\sum Y_i^2)$ beat $Y_i$ whenever $n > 2$; these estimates are not admissible. This estimate may be justified by noting that $P((n-2)/\sum Y_i^2) = 1/(\sigma^2 + \sigma_0^2)$ under the conditions of the theorem, so that the shrinking factor is estimated unbiasedly. The

Bayes estimates, in contrast, shrink rather less when $S$ is small than when it is large; for large $S$, the shrinking factor $P[V|S]$ will be close to $(n - \alpha)$; for small $S$, it will be close to zero.

If $(n - \alpha)$ is even, $P[V|S] = (n - \alpha)P[Z > (n - \alpha)/2]/P[Z \geq (n - \alpha)/2]$ where $Z$ is Poisson with expectation $\frac{1}{2}S/\sigma^2$. For example, for $n = 3$, $\alpha = 1$, $P[V|S] = 2[1 - e^{-(1/2)S/\sigma^2}(1 + \frac{1}{2}S/\sigma^2)]/[1 - e^{-(1/2)S/\sigma^2}]$, and the estimate is $Y_i[1/(1 - \exp(-\frac{1}{2}S/\sigma^2)) - 2\sigma^2/S]$.

Let $\sigma = 1$, and consider the sample $Y_1 = 1.2$, $Y_2 = -0.6$, $Y_3 = 0.8$. Then $S = 2.44$, the shrinking factor is .6, and the new estimates are .72, $-.36$, .48.

## 9.3.  Shrinking towards the Mean

Lindley and Smith (1972) use the prior $X_i \sim N(\theta_0, \sigma_0^2)$, independently for $i = 1, 2, \ldots, n$ given $\theta_0$, and $\theta_0 \sim N(0, \tau^2)$. For the moment $\sigma_0^2$ and $\tau^2$ will be assumed known. Then

$$P[X_i|\mathbf{Y}, \theta_0] = [Y_i/\sigma^2 + \theta_0/\sigma_0^2]/(1/\sigma^2 + 1/\sigma_0^2)$$

and    $Y_i \sim N[\theta_0, \sigma^2 + \sigma_0^2]$ independently for $i = 1, 2, \ldots, n$

$$\bar{Y}|\theta_0 \sim N\left(\theta_0, \frac{\sigma^2 + \sigma_0^2}{n}\right).$$

$$P[\theta_0|\mathbf{Y}] = P[\theta_0|\bar{Y}] = \frac{\sum Y_i}{\sigma^2 + \sigma_0^2}\left/\left(\frac{n}{\sigma^2 + \sigma_0^2} + \frac{1}{\tau^2}\right)\right.$$

$$P[X_i|\mathbf{Y}] = \left[Y_i/\sigma^2 + \bar{Y}\left(1 + \frac{\sigma_0^2 + \sigma^2}{n\tau^2}\right)\middle/\sigma_0^2\right]\left/\left(\frac{1}{\sigma^2} + \frac{1}{\sigma_0^2}\right)\right.$$

$$= [Y_i/\sigma^2 + \bar{Y}/\sigma_0^2]\left/\left(\frac{1}{\sigma^2} + \frac{1}{\sigma_0^2}\right)\right. \quad \text{if} \quad \tau = \infty.$$

$$P_{\mathbf{x}}\sum(P(X_i|\mathbf{Y}) - X_i)^2 = \sigma^2 + \left[(n - 1)\frac{1}{\sigma^2} + \frac{1}{\sigma_0^4}\sum(X_i - \bar{X})^2\right]\left(\frac{1}{\sigma^2} + \frac{1}{\sigma_0^2}\right)^2$$

$$< n\sigma^2 \quad \text{if} \quad \sum(X_i - \bar{X})^2/(n - 1) < 2\sigma_0^2 + \sigma^2.$$

Note that

$$P[X_i|\mathbf{Y}, \sigma_0^2] = \bar{Y} + (Y_i - \bar{Y})\left[1 - \frac{\sigma^2}{\sigma^2 + \sigma_0^2}\right].$$

If $\sigma_0^2$ has prior density $(\sigma^2 + \sigma_0^2)^{1 - ((1/2)\alpha)}$, $P[\sum(Y_i - \bar{Y})^2/(\sigma^2 + \sigma_0^2)] = P[\chi_{n-\alpha-1}^2|\chi_{n-\alpha-1}^2 < \sum(Y_i - \bar{Y})^2/\sigma^2]$ and the estimate beats $Y_i$ if $n \geq 5 - \alpha$, from Theorem 9.1. Shrinking towards the mean rather than towards an arbitrary constant loses a degree of freedom, but doesn't change the basic arguments.

## 9.4. A Random Sample of Means

Suppose $Y_i \sim N(X_i, 1)$ independently, and the $X_i$ are a random sample from some prior $P_0$. The $Y_i$ are then a random sample from the density $g$,

$$g(y) = P_0[\exp(-\tfrac{1}{2}(X - y)^2)/\sqrt{2\pi}].$$

Assuming first that $g$ is known, the posterior mean of $X_i$ given $Y_i$ is $P_0[X \exp(-\tfrac{1}{2}(X - Y_i)^2)]/P_0[\exp(-\tfrac{1}{2}(X - Y_i)^2)] = Y_i + (d/dY_i)\log g$.

If $g$ is not known, it is necessary to place a prior distribution on it so that the posterior expectation of the "correction" $(d/dY_i)\log g$ may be computed.

An "empirical Bayes" approach permits estimation of $g$ by any method, not necessarily a Bayesian method. It is known that $Y_1, \ldots, Y_n$ form a random sample from $g$. A density estimation technique might be used to estimate $(d/dY)\log g$. For example, if $\log g$ has a continuous first derivative at $y$,

$$P[Y - y\|Y - y| < \varepsilon]/P[(Y - y)^2\|Y - y| < \varepsilon] \to \frac{d}{dy}\log g \text{ as } \varepsilon \to 0.$$

Thus $(d/dy)\log g$ may be estimated by $\sum_{|Y_i - y| < \varepsilon}(Y_i - y)/\sum_{|Y_i - y| < \varepsilon}(Y_i - y)^2$ for sensibly selected $\varepsilon$. (For $\varepsilon$ large enough to include all data values, the estimate will be similar to the James–Stein estimate (9.2). The estimate $Y_i$ will be replaced by an estimate closer to the mean of those observations near $Y_i$.

## 9.5. When Most of the Means Are Small

In 9.2 and 9.3, $g$ is normal with mean 0 and unknown variance, and a prior distribution is placed on the variance. In many regression and analysis of variance problems, most of the means $X_i$ are very close to zero, but a few are quite large. Such a situation is not well represented by a normal $g$, because it is not sufficiently long tailed. One alternative is to assume that $X_i$ comes from a distribution $p\delta_0 + (1 - p)N(0, \sigma_0^2)$ where $\delta_0\{0\} = 1$. Then $Y_i$ is a random sample from $pN(0, 1) + (1 - p)N(0, \sigma_0^2 + 1)$,

$$g(y) = \frac{1}{\sqrt{2\pi}}\left\{ p\exp(-\tfrac{1}{2}y^2) + \frac{(1 - p)}{\sqrt{1 + \sigma_0^2}}\exp(-\tfrac{1}{2}y^2/(1 + \sigma_0^2)) \right\}$$

$$\frac{d}{dy}\log g(y) = \frac{-y\left\{ p\exp(-\tfrac{1}{2}y^2) + \frac{(1 - p)}{(1 + \sigma_0^2)^{3/2}}\exp(-\tfrac{1}{2}y^2/(1 + \sigma_0^2)) \right\}}{\left\{ p\exp(-\tfrac{1}{2}y^2) + \frac{1 - p}{\sqrt{1 + \sigma_0^2}}\exp(-\tfrac{1}{2}y^2/(1 + \sigma_0^2)) \right\}}$$

If $y$ is small, the adjustment is close to $-y$; if $y$ is large it is close to $y/(1 + \sigma_0^2)$; in this way the small observed values $Y_i$ are moved very close to zero, but the large observed values $Y_i$ are relatively unchanged. In practice $p$ and $\sigma_0^2$ must be estimated from the $Y_i$. A Bayesian approach requires computation of the posterior mean of $(d/dy) \log g(y)$ but no prior on $p$ and $\sigma_0$ is known which permits explicit computation. It is straightforward computationally to estimate $p$ and $\sigma_0^2$ to maximize the likelihood of the observations, but explicit expressions are not available, and it is not known whether the resulting estimates of the $X_i$ beat the straight estimates.

A standard approach to the problem of many small means is to carry out a significance test on each mean separately, and to set all those means to zero which do not exceed some significance level. Here, the estimate would be $\hat{X}_i = Y_i\{|Y_i| \geq c\}$, where $c$ is the cutoff point in the significance test. Then

$$\sum P(Y_i - X_i)^2 - \sum P(\hat{X}_i - X_i)^2 = \sum P(\{|Y_i| < c\}(Y_i^2 - 2Y_iX_i))$$
$$= \sum P(\{|Z_i + X_i| < c\}(Z_i^2 - X_i^2))$$

where $Z_i \sim N(0, 1)$. If $|X| > 1$, $P(\{|Z + X| < c\}(Z^2 - X^2)) < 0$ for every choice of $c$. Thus there is no way to choose $c$ so that the estimates $\hat{X}_i$ have uniformly smaller mean square error than $Y_i$; it does not help to allow $c$ to depend on the $Y_i$.

Yet there is practical value in setting many small means to be exactly zero if there is no evidence of significant departure from zero. Suppose the loss function is

$$L(d, s) = \{d \neq s\} + k(d - s)^2.$$

Let $P_0$ be a unitary probability on $S$ which has an atom $P_0\{s_0\}$ only at $s_0$. Then the probable loss for $d$ is $P_0\{d \neq s\} + kP_0(d - s)^2 = P_0 s_0 + \{d = s_0\}(1 - 2P_0 s_0) + k(d - P_0 s)^2 + kP_0(s - P_0 s)^2$. The Bayes decision is $d = s_0$ if $2P_0\{s_0\} > k(s_0 - P_0 s)^2 + 1$, and $d = P_0 s$ otherwise.

If $Y_i \sim N(X_i, 1)$ independently, where the $X_i$ are sampled from $p\delta_0 + (1 - p)N(0, \sigma_0^2)$, then

$$X_i| Y_i \sim p_{Y_i}\delta_0 + (1 - p_{Y_i})N\left[\frac{Y_i}{1 + \dfrac{1}{\sigma_0^2}}, \frac{1}{1 + \dfrac{1}{\sigma_0^2}}\right]$$

where

$$p_{Y_i} = \left\{1 + \frac{1 - p}{p}\cdot\frac{1}{(1 + \sigma_0^2)^{1/2}} \exp -\left[\tfrac{1}{2}y^2 / \left(\frac{1}{\sigma_0^2} + 1\right)\right]\right\}^{-1}$$

is the posterior probability that $Y_i$ came from the 0 component. The Bayes estimate will be $X_i = 0$ if

$$1 + \left[(1 - p_{Y_i})\frac{Y_i}{\dfrac{1}{\sigma_0^2} + 1}\right]^2 k < 2p_{Y_i},$$

and

$$X_i = (1 - p_{Y_i}) \frac{Y_i}{\dfrac{1}{\sigma_0^2} + 1}$$

otherwise.

## 9.6. Multivariate Means

Let $Y \sim N(X, \Sigma)$, $X \sim N(0, k\Sigma_0)$ where $Y$ and $X$ are $n$ dimensional vectors, $\Sigma$ and $\Sigma_0$ are known covariance matrices, and $k$ is unknown. By a linear transformation applied to $Y$ and $X$, this case may be reduced to $Y_i \sim N(X_i, \sigma_i^2)$, $X_i \sim N(0, \sigma_0^2)$ where $\sigma_0^2$ is unknown, and the distributions are independent for different $i$. See Efron and Morris (1973).

Given $\sigma_0^2$, $P(X_i | Y_i) = Y_i(1 - (\sigma_i^2/\sigma_0^2 + \sigma_i^2))$.

A Bayes procedure for a prior density $f(\sigma_0^2)$ on $\sigma_0^2$ would use

$$P\left[ \left( \frac{1}{\sigma_0^2 + \sigma_i^2} \right) \middle| Y \right] = \frac{\int \frac{1}{\sigma_0^2 + \sigma_i^2} \cdot \prod (\sigma_0^2 + \sigma_i^2)^{-1/2} \exp(-\tfrac{1}{2} Y_i^2/(\sigma_0^2 + \sigma_i^2)) f \, d\sigma_0^2}{\int \prod (\sigma_0^2 + \sigma_i^2)^{-1/2} \exp[-\tfrac{1}{2} \sum Y_i^2/(\sigma_0^2 + \sigma_i^2)] f \, d\sigma_0^2}$$

but no magical $f$ exists that permits a simple explicit computation. As a practical matter, taking a uniform discrete prior on $\sigma_0^2$ from 0 to $\sum Y_i^2(1 + \sqrt{8/n})$ in 100 steps, should give a reasonable Bayes estimate of $1/(\sigma_0^2 + \sigma_i^2)$. [For a continuous $f$, the above integrals will have to be approximated as if the prior were discrete, anyway.]

A simple alternative to a Bayes procedure uses $P[Y_i^2 | \sigma_0^2] = \sigma_0^2 + \sigma_i^2$, $P[\sum Y_i^2 | \sigma_0^2] = n\sigma_0^2 + \sum \sigma_i^2$, so that $\sigma_0^2$ is estimated unbiasedly by $\sum (Y_i^2 - \sigma_i^2)/n$. This estimate is sometimes embarrassed by being negative, and may not lead to a good estimate of $1/(\sigma_0^2 + \sigma_i^2)$.

A slightly better non-Bayesian method is maximum likelihood, which finds $\sigma_0^2$ to maximize $-\sum \log(\sigma_i^2 + \sigma_0^2) - \sum Y_i^2/(\sigma_i^2 + \sigma_0^2)$. The maximum value occurs at 0, or at a solution of an equation $\sum (Y_i^2 - \sigma_i^2 - \sigma_0^2)/(\sigma_i^2 + \sigma_0^2)^2 = 0$; thus $\sigma_0^2$ is the weighted average of $Y_i^2 - \sigma_i^2$ with weights inversely proportional to the variances of $Y_i^2$ given $\sigma_0^2$. However the solution may not be unique, and checking a spectrum of $\sigma_0^2$ values is about as difficult as doing a Bayes approximate integration.

The above procedures are not known to be uniformly better than the straight estimates $Y_i$, which have sum of squared error loss $\sum \sigma_i^2$. Given $\sigma_0^2$, the loss of the Bayes estimate is $\sum (\sigma_i^2(\sigma_0^4/(\sigma_0^2 + \sigma_i^2)^2) + \sigma_i^4 X_i^2/(\sigma_0^2 + \sigma_i^2)^2)$ which is less than $\sum \sigma_i^2$ if

$$\sum (\sigma_0^{-2} + \sigma_i^{-2})^{-2}[-2\sigma_0^2 - \sigma_i^2 + X_i^2] < 0.$$

This condition is analogous to one given in 9.3; it will always be satisfied for $\sigma_0^2$ large enough. This suggest that an estimate beating $Y_i$ might be obtained by overestimating $\sigma_0^2$. Of course, if loss is measured by $P(\sum(\hat{X}_i - X_i)^2 / \sigma_i^2 | \mathbf{X})$, the problem may be transformed to one in which all the $\sigma_i^2$ are equal and Stein's estimate and unitary Bayes estimates exist beating $Y_i$. Brown (1966) shows the estimate $Y_i$ to be inadmissible for a large class of loss functions; better estimators are given by Brandwein and Strawderman (1978).

## 9.7. Regression

Suppose $Y|X \sim N(AX, \sigma^2 I_n)$, $X \sim N(0, \sigma_0^2 I_p)$ where $Y$ is $n \times 1$, $A$ is $n \times p$, $X$ is $p \times 1$ and $I_n$ denotes an $n \times n$ unit matrix. Then

$$X|Y \sim N[(A'A/\sigma^2 + I/\sigma_0^2)^{-1}A'Y/\sigma^2, (A'A/\sigma^2 + I/\sigma_0^2)^{-1}].$$

The estimate $P(X|Y)$ of $X$ is often advocated for purely computational reasons, to guard against singularity or near-singularity of $A'A$, Hoerl and Kennard (1970). As in 9.6, it is difficult to estimate $\sigma_0^2$ by a simple Bayes procedure. It is tempting to use the unbiased estimate

$$\hat{\sigma}_0^2 = [Y'A(A'A)^{-1}A'Y - p\sigma^2]/\text{trace}(A'A),$$

but this is dangerous because it might be negative.

The maximum likelihood estimate for $\sigma_0^2$, assuming $\sigma^2$ known, minimizes $\log|\sigma_0^2 AA' + \sigma^2 I| + Y'(\sigma_0^2 AA' + \sigma^2 I)^{-1}Y$. This looks nasty, but when $AA'$ is diagonalized, it reduces to the likelihood expression in 9.6.

See Lindley and Smith (1972).

## 9.8. Many Means, Unknown Variance

Let $Y_i|X_i \sim N(0, \sigma^2)$, $X_i \sim N(0, \sigma_0^2)$, $i = 1, 2, \ldots, n$, independently for each $i$, and suppose there is an independent estimate of variance $S$, with $S \sim \sigma^2 \chi_k^2$. Such a situation arises in regression problems. Given $\sigma^2$ and $\sigma_0^2$,

$$P[X|Y] = Y_i\left[1 - \frac{\sigma^2}{\sigma^2 + \sigma_0^2}\right].$$

The density of $S$, $\mathbf{Y}$ given $\sigma^2$ and $\sigma_0^2$ is proportional to

$$\frac{S^{(k/2)-1}}{(\sigma^2)^{k/2}} \exp(-\tfrac{1}{2}S/\sigma^2) \frac{(\sum Y_i^2)^{(n/2)-1}}{(\sigma^2 + \sigma_0^2)^{n/2}} \exp\left[-\tfrac{1}{2}Y_i^2/(\sigma^2 + \sigma_0^2)\right]$$

We may estimate $\sigma^2$ and $\sigma_0^2$ unbiasedly by solving

$$S = k\sigma^2, \qquad \sum_i Y_i^2 = n(\sigma^2 + \sigma_0^2),$$

but it is better to estimate the coefficient $\sigma^2/(\sigma_i^2 + \sigma_0^2)$ unbiasedly by $(n-2)/k \cdot S/\sum Y_i^2$; the estimator $Y_i[1 - ((n-2)/k) \cdot S/\sum Y_i^2]$ beats $Y_i$, from Baranchik (1971). Even so the estimator can occasionally give foolish results with the coefficient of $Y_i$ negative.

A maximum likelihood procedure gives the same results as the unbiased method $S = k\sigma^2$, $\sum Y_i^2 = n(\sigma^2 + \sigma_0^2)$ except when $\sigma_0^2$ is estimated negative; in that case $\sigma_0^2$ is estimated to be zero, and $\sigma^2$ is estimated by $(S + \sum Y_i^2)/(n+k)$.

For the prior density $(\sigma^2 + \sigma_0^2)^{((1/2)\alpha)-1}(\sigma^2)^{((1/2)\beta)-1}$, from 9.2, $\sigma^{-2} \sim \chi_{k-\beta}^2/S$, $(\sigma^2 + \sigma_0^2)^{-1} \sim \chi_{n-\alpha}^2/\sum Y_i^2$ where the $\chi_{k-\beta}^2$, $\chi_{n-\alpha}^2$ are sampled from independent chi-squares but accepted only if $\sigma^{-2} \geq (\sigma^2 + \sigma_0^2)^{-1}$. Thus $\sigma^2/(\sigma^2 + \sigma_0^2) \sim (S/\sum Y_i^2)\chi_{n-\alpha}^2/\chi_{k-\beta}^2$ constrained not to exceed 1, and $P[\sigma^2/(\sigma^2 + \sigma_0^2)|\mathbf{Y}] = (S/\sum Y_i^2)P[\chi_{n-\alpha}^2/\chi_{k-\beta}^2|\chi_{n-\alpha}^2/\chi_{k-\beta}^2 \leq \sum Y_i^2/S]$. The computation is an incomplete beta integral. From Baranchik (1971), the estimator $Y_i[1 - (S/\sum Y_i^2)r(\sum Y_i^2/S)]$ beats $Y_i$ if $r$ is non-decreasing, $r \leq 2(n-2)/(k+2)$. Here $r$ is obviously non-decreasing, $r \leq P[\chi_{n-\alpha}^2/\chi_{k-\beta}^2] = (n-\alpha)/(k-\beta-2)$. Thus the posterior mean beats $Y_i$ if $(n-\alpha)/(k-\beta-2) \leq 2(n-2)/(k+2)$. (These estimates are not Bayes because the loss is not integrable. For example, when $\alpha = 0$, $\beta = 0$ the condition is satisfied for no $k, n$; when $\alpha = 2$, $\beta = -4$, it is satisfied for $n \geq 3$.

# 9.9. Variance Components, One Way Analysis of Variance

Suppose that a number of normal samples estimate the means $X_1, \ldots, X_n$; for the $j^{th}$ sample

$$Y_{ij} \sim N(X_j, \sigma^2), \qquad i = 1, \ldots, m$$

The $X_j$'s are assumed to be sampled from $N(X, \sigma_0^2)$. Finally $X \sim N(0, \sigma_M^2)$. Since $\bar{Y}_j \sim N[X_j, \sigma^2/m]$ this is essentially the same situation considered in 9.3. Given $\sigma_0^2, \sigma^2, \sigma_M^2$ there will be posterior mean estimates of the $X_j$. In practice, it is necessary to estimate the "variance components" $\sigma_0^2, \sigma^2, \sigma_M^2$ somehow, and they are of interest in themselves to indicate how important between group effects (represented by $\sigma_0^2$) and within group effects (represented by the $\sigma_j^2$) are:

Here

$$\sum_j \sum_i (Y_{ij} - \bar{Y}_j)^2 \sim \sigma^2 \chi_{n(m-1)}^2$$

$$\sum (Y_j - \bar{Y})^2 \sim \left(\frac{\sigma^2}{m} + \sigma_0^2\right)\chi_{n-1}^2$$

$$\bar{Y}^2 \sim \left(\frac{\sigma^2}{mn} + \frac{\sigma_0^2}{n} + \sigma_M^2\right)\chi_1^2$$

independently. (The distributions are not so simple if there are unequal numbers in the different samples.) Unbiased estimates of $\sigma^2$, $\sigma_0^2$ and $\sigma_M^2$ may be easily constructed from linear combinations of the sums of squares in $Y$, but the estimates are inadmissible because they may be negative. Maximum likelihood gives the same estimates if the solutions to the equations are positive.

For a prior uniform in $\log \sigma^2$, $\log(\sigma^2/m + \sigma^2)$ and $\log(\sigma^2/mn + \sigma_0^2/n + \sigma_M^2)$, the posterior distribution of $\sigma^2$, $\sigma^2/m + \sigma_0^2$, $\sigma^2/mn + \sigma_0^2/n + \sigma_M^2$ is $\sum\sum(Y_{ij} - \bar{Y}_j)^2/\chi^2_{n(m-1)}$, $\sum(\bar{Y}_j - \bar{Y})^2/\chi^2_{n-1}$, $\bar{Y}^2/\chi^2_1$ where the chi-squares are taken independently, but accepted only if the appropriate inequalities hold between the three variables. Computation of posterior means would require formidable numerical integrations in three dimensions. Similar considerations arise in estimating variance components for more complicated analysis of variance models.

## 9.10. Problems

P1. A surveyor, poor but honest, measures the three angles $(\theta_1, \theta_2, \theta_3)$ of a triangle with independent errors $N(0, 1)$. The measured angles are $\hat{\theta}_1 = 63°$, $\hat{\theta}_2 = 31°$, $\hat{\theta}_3 = 92°$. For a suitable prior on $\boldsymbol{\theta}$, find the posterior distributions of each of $\theta_1, \theta_2, \theta_3$ given the data. [The true values should add to 180°.]

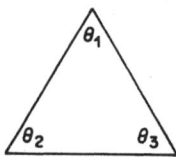

P2. In football, the scoring difference between team $i$ and team $j$ is distributed as $N[\mu_i - \mu_j, \sigma^2]$. The prior distributions at the beginning of a season are, independently,

| | |
|---|---|
| Yale | $\mu_1 \sim N(0, \sigma^2)$ |
| Harvard | $\mu_2 \sim N(0, \sigma^2)$ |
| Princeton | $\mu_3 \sim N(0, \sigma^2)$ |
| Dartmouth | $\mu_4 \sim N(6, \sigma^2)$ |

Game scores are: Harvard 13–Princeton 6
                 Princeton 27–Dartmouth 20
                 Princeton 21–Yale 3.

Compute the probability that Harvard will beat Yale, given the observed scores.

P3. For $Y_i \sim N(X_i, 1)$   independent
        $X_i \sim N(X_0, \sigma_0^2)$   independent
    assume $g(X_0, \sigma_0^2) = \cdot 1/(\sigma_0^2 + 1)$.
        Find the posterior mean of $X_i$ given $Y_1, \ldots, Y_n$.

P4. Votes for the Democratic candidate for President:

| South | Central | New England |
|-------|---------|-------------|
| 21 | 43 | 61 |
| 29 | 47 | 62 |
| 30 | 42 | 65 |
| 27 | | |
| 21 | | |

Construct a model II analysis of variance, and estimate variance components, using unbiased estimates and Bayes estimates.

P5. Show that the following estimate in the Stein problem, $Y_i \sim N(\theta_i, 1)$, is Bayes and beats $Y_i$:

$$\hat{\theta}_i = Y_i \left\{ 1 - \frac{2}{\sum Y_i^2} \frac{[1 - e^{-(1/2)\Sigma Y_i^2}(1 + \frac{1}{2}\sum Y_i^2)]}{1 - e^{-(1/2)\Sigma Y_i^2}} \right\}$$

Show that the multiplier is never negative.

# 9.11. References

Baranchik, A. J. (1970), A family of minimax estimators of the mean of a multivariate normal distribution, *Ann. Math. Statist.* **41**, 642–645.

Brandwein, A. R. and Strawderman, W. E. (1978), Minimax estimation of location parameters for spherically symmetric unimodal distributions under quadratic loss, *Annals of Statistics* **6**, 377–416.

Brown, L. D. (1966), On the admissibility of invariant estimators of one or more location parameters, *Ann. Math. Statist.* **37**, 1083–1136.

Efron, B. and Morris, C. (1973), Stein's estimation rule and its competitors—an empirical Bayes approach, *J. Am. Stat. Ass.* **68**, 117–130.

Hoerl, A. E. and Kennard, R. W. (1970), Ridge regression: biased estimation for non-orthogonal problems, *Technometrics* **12**, 69–82.

James, W. and Stein, C. (1961), Estimation with quadratic loss, *Proc. Fourth Berkeley Symposium, University of California Press*, **1**, 361–379.

Jeffreys, H. (1961), *Theory of Probability*. Cambridge University Press, Cambridge.

Lindley, D. V. and Smith A. F. M. (1972), Bayes estimates for the linear model, *J. Roy. Stat. Soc.* B **34**, 1–41.

Stein, C. (1956), Inadmissibility of the usual estimator for the mean of a multivariate normal population, *Proc. Third Berkeley Symposium* **1**, 197–206.

Strawderman, W. (1971), Proper Bayes minimax estimators of the multivariate normal mean, *Ann. Math. Statist.* **42**, 385–388.

# CHAPTER 10

# The Multinomial Distribution

## 10.0. Introduction

A discrete random variable $X$ takes values $i = 1, 2, \ldots, k$ with probabilities $\{p_i, i = 1, 2, \ldots, k\}$. A sample of size $n$ from $X$ gives the value $X = i$ $n_i$ times. The multivariate distribution $\{n_i, i = 1, \ldots, k\}$ is *multinomial* with parameters $n$, $\{p_i, i = 1, \ldots, k\}$. It is ubiquitous in problems dealing with discrete data. The values $1, 2, \ldots, k$ are called categories or cells.

If $(n_1, n_2, \ldots, n_k)$ is multinomial $n$, $\{p_i\}$ then $n_1 + n_2, n_3, \ldots, n_k$ is multinomial $n$, $(p_1 + p_2, p_3, \ldots, p_k)$; $n_1, n_2, \ldots, n_j$ given $\sum_{i=1}^{j} n_i$ is multinomial $\sum_{i=1}^{j} n_i$, $\{p_i / \sum_{i=1}^{j} p_i, i = 1, \ldots, j\}$.

The multinomial is obtained from $k$ independent Poissons $n_i$ with expectations $\lambda_i$; the distribution of $n_1, \ldots, n_k$ given $n = \sum n_i$ is multinomial with parameters $n$, $\{\lambda_i / \sum_{i=1}^{k} \lambda_i, i = 1, \ldots, k\}$. This fact is very convenient in formulating models and handling computations, because the Poisson $n_i$ are independent.

In general, the interesting problems in asymptotics and decision theory arise when some of the $p_i$ are small. For example the usual maximum likelihood estimates of $p_i$ are inadmissible if $|p_i| \geq \varepsilon$, $i = 1, \ldots, k$.

Standard families of prior distributions exist for the multinomial, but they don't work too well for many parameter problems. It is necessary to incorporate expected similarities between the $p_i$'s into the prior for many parameter problems.

## 10.1. Dirichlet Priors

The multinomial $X$ taking values $i = 1, 2, \ldots, k$ with probabilities $p_i$

$$p(X) = \prod p_i^{\{X = i\}}$$

$$= \exp\left[ \sum_{i=1}^{k-1} \{X = i\} \log[p_i/(1 - p_k)] + \log(1 - p_k) \right]$$

is exponential $E[\mu, \mathbf{T}, \boldsymbol{\theta}]$ where $\mathbf{T}$ is the vector $\{X = i\}$, $i = 1, \dots, k - 1$ and $\boldsymbol{\theta}$ is the vector $\log[p_i/(1 - p_k)]$, $i = 1, \dots, k - 1$, and $\mu$ is counting measure on $1, 2, \dots, k$. Because of the asymmetry of this parameterization, it is often convenient to think of the multinomial as $E\{\mu, [\{X = i\}, i = 1, \dots, k],$ $[\log p_i, i = 1, \dots, k]\}$ where the parameters $\log p_i$ are constrained to lie in a $(k - 1)$ dimensional subset $(\sum p_i = 1)$ of $R^k$. For $n$ observations, with $n_i = \sum_{j=1}^{n}\{X_j = i\}$,

$$p[X_1, \dots, X_n] = \prod p_i^{n_i}$$

$$p[n_1, \dots, n_k] = \frac{n!}{\prod n_i!}\prod p_i^{n_i}.$$

If $\{p_i\}$ is uniformly distributed over the simplex $p_i \geq 0$, $\sum p_i = 1$, the posterior density given $n_1, \dots, n_k$ is $\prod p_i^{n_i}((n + k - 1)!/\prod n_i!)$. More generally, the *Dirichlet density* on $\{p_i\}$, with respect to the uniform $\mu$ over $p_i \geq 0$, $\sum p_i = 1$ is $d_\alpha(\mathbf{p}) = \Gamma(\sum \alpha_i)\prod p_i^{\alpha_i - 1}/\Gamma(\alpha_i)$; the Dirichlet probability is $D_\alpha = E[\mu, \{\log p_i\}, \{\alpha_i - 1\}]$.

The Dirichlet generalizes the beta to many dimensions. If $U_i$ are independent gamma with densities $\propto u_i^{\alpha_i - 1}\exp(-au_i)$, then $\{U_i/\sum_{i=1}^{k} U_i\}$ is Dirichlet $D_\alpha$ (similarly to the multinomial being independent Poissons $n_1, \dots, n_k$ conditioned by $\sum n_i = n$). If the prior density is $d_\alpha$, then the posterior given $n_1, \dots, n_k$ is $d_{\alpha + \mathbf{n}}$.

$$D_\alpha[\mathbf{P}] = \alpha/\sum \alpha_i, \quad D_\alpha[\mathbf{PP'}] = \left(\alpha\alpha' + \begin{bmatrix} \alpha_1 & & \\ & \ddots & \\ & & \alpha_k \end{bmatrix}\right)\Big/\sum \alpha_i(\sum \alpha_i + 1).$$

If $(p_1, \dots, p_k)$ is Dirichlet $D_\alpha$ then $p_1, p_2, \dots, p_r, 1 - \sum_{i=1}^{r} p_i$ is $D_{\alpha_1, \dots, \alpha_r, \sum_{i>r}\alpha_i}$, and $p_1, \dots, p_r$ given $p_{r+1}, \dots, p_k$ is $(1 - \sum_{i>r} p_i)D_{\alpha_1, \dots, \alpha_r}$.

## 10.2. Admissibility of Maximum Likelihood, Multinomial Case

The maximum likelihood estimates of $p_i$ are $n_i/n$; these are posterior means for the non-unitary prior density $1/\prod p_i$. They are the only estimates that do not depend on the fineness of subdivision of the multinomial cells, Johnson (1932). If there are very many $p_i$, all probability estimates will be $0/n$ or $1/n$ which is unsatisfactory.

**Theorem** (Johnson (1971)). *The maximum likelihood estimator $\hat{p}_i = n_i/n$ is an admissible estimator of $p_i$ with loss function* $L(\mathbf{d}, \mathbf{p}) = \sum_{i=1}^{k}(p_i - d_i)^2$.

PROOF. The technique of 6.3 would approximate $\hat{p}_i$ by Bayes estimates for densities $\prod p_i^{\alpha - 1}$, with $\alpha \to 0$, but this is not effective for $k > 2$ because more than one of the $n_i$ may be zero, and this causes irretrievable degeneracy in the posterior densities when the corresponding $p_i$ are zero. The essence of Johnson's proof is careful handling of the cases where some $n_i$ are zero.

Consider first $k = 2$ and suppose $\delta$ has risk nowhere greater than the risk of $n_i/n$.

$$r(\delta, \mathbf{p}) = \sum_{n_1 + n_2 = n} \binom{n}{n_1} [(\delta_1 - p_1)^2 + (\delta_2 - p_2)^2] p_1^{n_1} p_2^{n_2}$$
$$= 0 \text{ at } p_1 = 0 \text{ or } p_2 = 0$$

since $n_i/n$ has zero risk for $p_1 p_2 = 0$. Thus $\delta_1(0, n) = 0$, $\delta_2(0, n) = 1$ and so $\delta$ and $\mathbf{n}/n$ agree when $n_1 = 0$ or $n_1 = n$.

$$r(\delta, \mathbf{p}) - r(\mathbf{n}/n, \mathbf{p}) = \sum_{0 < n_1 < n} \binom{n}{n_1} \sum_{i=1}^{2} [(\delta_i - p_i)^2 - (n_i/n - p_i)^2] p_1^{n_1} p_2^{n_2}$$

$$\int [r(\delta, \mathbf{p}) - r(\mathbf{n}/n, \mathbf{p})] \frac{dp_1}{p_1 p_2}$$

$$= \sum_{0 < n_1 < n} \binom{n}{n_1} \int \sum_{i=1}^{2} [(\delta_i - p_i)^2 - (n_i/n - p_i)^2] p_1^{n_1 - 1} p_2^{n_2 - 1} dp_1$$

$$\geq 0 \text{ since } n_i/n \text{ is Bayes on prior } 1/p_1 p_2,$$

with equality only if $\delta_i = n_i/n$, $i = 1, 2$. Thus if $\delta$ has risk no greater than $\mathbf{n}/n$, it equals $\mathbf{n}/n$; thus $\mathbf{n}/n$ is admissible.

Note that integration is possible after multiplying by $1/p_1 p_2$ because the cases $n_1 = 0$ and $n_1 = n$ have been eliminated.

Consider next $k = 3$, and suppose that $\delta$ has risk no greater than that of $\mathbf{n}/n$. Letting $p_1 = 0$,

$$r(\delta, \mathbf{p}) = \sum_{\mathbf{n}} \binom{n}{\mathbf{n}} \sum_{i=1}^{3} (\delta_i - p_i)^2 \prod p_i^{n_i}$$

$$= \sum_{n_2 + n_3 = n} \binom{n}{n_2} \sum_{i=2}^{3} (\delta_i(0, n_2, n_3) - p_i)^2 \prod_{i=2,3} p_i^{n_i}$$

$$+ \sum_{n_2 + n_3 = n} \binom{n}{n_2} \delta_1^2(0, n_2, n_3)) \prod_{i=2,3} p_i^{n_i}$$

Since $r(\delta, \mathbf{p}) \leq r(\mathbf{n}/n, \mathbf{p})$, then $\sum_{n_2 + n_3 = n} \binom{n}{n_2} \sum_{i=2}^{3} (\delta_i(0, n_2, n_3) - p_i)^2 \prod_{i=2,3} p_i^{n_i}$ is not greater than $r(0, n_2/n, n_3/n; \mathbf{p})$, which implies $\delta_i(0, n_2, n_3) = n_i/n$, $i = 1, 2, 3$. [Note that $\delta_1(0, n_2, n_3) = 0$ since otherwise, for $p_1 = 0$, $r(\delta, \mathbf{p}) > r(\mathbf{n}/n, \mathbf{p})$.] Similarly, $\delta$ agrees with $\mathbf{n}/n$ whenever $n_1$ or $n_2$ or $n_3 = 0$.

$$\int [r(\delta, \mathbf{p}) - r(\mathbf{n}/n, \mathbf{p})] \frac{dp_1 dp_2}{p_1 p_2 p_3} = \sum_{\mathbf{n}, n_i > 0} \binom{n}{\mathbf{n}} \int \sum [(\delta_i - p_i)^2$$
$$- (n_i/n - p_i)^2] \prod p_i^{n_i - 1}$$
$$\geq 0,$$

since $n_i/n$ is Bayes for the density $1/\prod p_i$ with equality only if $\delta_i = n_i/n$. The integration is justified because $n_i > 0$. Thus if $r(\delta, \mathbf{p}) \leq r(\mathbf{n}/n, \mathbf{p})$, $\delta = \mathbf{n}/n$, so $\mathbf{n}/n$ is admissible.

General $k$ is handled by induction; if one of the $p$'s is set zero, the decision procedure with the corresponding $n_i$ zero must coincide with maximum likelihood; so the difference between two procedures need only be assessed over $n_i > 0$; and integrating with respect to $1/\prod p_i$ shows that $\delta$ cannot beat $n$.

[Decisions of form: take $\delta_j$ with probability $d_j$ have risk exceeding that of $\sum d_j \delta_j$ so they can't beat $\mathbf{n}/n$ either.]                    □


## 10.3. Inadmissibility of Maximum Likelihood, Poisson Case

If the $n_i$ are independent Poissons with expectations $\lambda_i$, then $n_1, \ldots, n_k$ given $\sum n_i = n$ are multinomial with parameters $\lambda_i/\sum \lambda_i$. For this reason, it is often convenient to formulate multinomial models using independent Poissons. A convenient family of prior densities for the Poisson $P_\lambda$, $P_\lambda\{n\} = e^{-\lambda}\lambda^n/n!$ is the gamma $G(m, a)$ with density $a^m \lambda^{m-1} e^{-a\lambda}/\Gamma(m)$; given observation $n$, the posterior density is $G(m + n, a + 1)$.

**Theorem** (Clevenson and Zidek (1975)). *Let $n_i$ be independent Poisson with expectations $\lambda_i$, $i = 1, 2, \ldots, k$. Then, for $k \geq 2$*

$$P\left(\sum\left[n_i \Big/ \left(1 + \frac{k-1}{\sum n_i}\right) - \lambda_i\right]^2 \Big/ \lambda_i\right) < P(\sum(n_i - \lambda_i)^2/\lambda_i)$$

*for all $\lambda_i$. Thus $n_i$ is inadmissible as an estimate of $\lambda_i$.*

PROOF. Given $n = \sum_{i=1}^k n_i$, the $n_i$ are multinomial with parameters $p_i = \lambda_i/\sum \lambda_i$

$$P[\sum(an_i - \lambda_i)^2/\lambda_i \mid n] = \sum[a^2 np_i(1 - p_i)/\lambda_i + (anp_i - \lambda_i)^2/\lambda_i]$$
$$= a^2 n(k - 1)/\sum \lambda_i + (an - \sum \lambda_i)^2/\sum \lambda_i$$

The value $a$ which minimizes this expression is $\sum \lambda_i/(n + k + 1)$, which is estimated by $n/(n + k + 1) = a_n$. Set $\lambda = \sum \lambda_i$.

$$P[\sum(a_n n_i - \lambda_i)^2/\lambda_i] = a_n^2(n(k - 1) + n^2)/\sum \lambda_i - 2a_n n + \sum \lambda_i$$
$$= n^3/(n + k - 1)\sum \lambda_i - 2n^2/(n + k - 1) + \sum \lambda_i$$

$$n^3/(n + k - 1) - 2n^2\lambda/(n + k - 1) = n^2 - (2\lambda + k - 1)n + (k - 1)(2\lambda + k + 1)$$
$$- \frac{(k - 1)^2(2\lambda + k - 1)}{n + k - 1}$$

$$P[(n^3 - 2n^2\lambda)/(n + k - 1)] < \lambda^2 + \lambda + (2\lambda + k - 1)(k - 1 - \lambda)$$
$$- (k - 1)^2(2\lambda + k - 1)/(\lambda + k - 1)$$

since $X$ and $1/X$ are negatively correlated.

$$P[\sum (a_n n_i - \lambda_i)^2/\lambda_i] < 2\lambda + 1 - (2\lambda + k - 1)\lambda/(\lambda + k - 1)$$
$$< k - (k-1)^2/(\lambda + k - 1) < k = P[\sum (n_i - \lambda_i)^2/\lambda_i]. \quad \square$$

## 10.4. Selection of Dirichlet Priors

Jeffreys's prior is $D_{1/2,1/2,\ldots,1/2}$ giving density $\propto \prod p_i^{-1/2}$, and posterior means $(n_i + 1/2)/(\sum n_i + k/2)$; the cell estimate depends significantly on $k$, so that if other cells are subdivided into, say, 100 more cells, a given estimate is substantially reduced. Perks (1947) suggests $\prod p_i^{-1/k}$ which gives estimates $(n_i + 1/k)/(\sum n_i + 1)$. Following the binomial case, it is useful to consider the family of prior densities $p_i^{\alpha_i - 1}$, $\alpha_i > 0$, $\sum \alpha_i = 1$ which gives ranges of estimates unaffected by amalgamation or subdivision of cells; these are analogous to confidence priors in the binomial case.

Another possibility is to estimate the Dirichlet prior; assume the prior $\prod p_i^{\alpha - 1}$; then $P[n_i] = n/k$, $P[n_i^2] = n/k + (n-1)(\alpha(\alpha + 1)n/k\alpha(k\alpha + 1))$,

$$P(\sum n_i^2) = n + n(n-1)(\alpha + 1)/(k\alpha + 1)$$

Thus $\alpha$ is estimated by solving $\sum n_i^2 = n + n(n-1)(\alpha + 1)/(k\alpha + 1)$, or equivalently, $(1/(k-1))\sum (n_i - n/k)^2 = n(k\alpha + n)/(k\alpha + 1)k$. There may be no non-negative solution $\alpha$ if $n_i$ has small enough variance; set $\alpha = \infty$ if $(1/(k-1))\sum (n_i - n/k)^2 \leq n/k$.

A more satisfactory (and more difficult) procedure due to Good (1965) selects $\alpha$ to maximize the likelihood

$$P(\mathbf{n}|\alpha) = \frac{\Gamma(k\alpha)\Gamma(n+1)}{\Gamma(\alpha)^k \Gamma(n+\alpha k)} \prod \frac{\Gamma(n_i + \alpha)}{\Gamma(n_i + 1)}.$$

Good (1975) shows that $P(\mathbf{n}|\alpha)$ is maximized by $\alpha = \infty$ when the chi-square goodness of fit statistic, $X = (k/n)\sum (n_i - n/k)^2 \leq k - 1$. He suggests using $G = \sup_\alpha [2 \log P(\mathbf{n}|\alpha)/P(\mathbf{n}_0|\alpha)]^{1/2}$ as a test statistic for deciding $p_i = 1/k$, where $n_{0i} = n/k$. In Good (1967) it is asserted that $G^2$ is distributed as $\chi_1^2$ given that $G > 0$, asymptotically as $n \to \infty$ for $k$ fixed. However, asymptotically, the expansions for gamma functions (Abramowitz and Stegun, 1964, p. 257) show

$$G(\alpha) = \log P(\mathbf{n}|\alpha)/P(\mathbf{n}_0|\alpha) \approx \tfrac{1}{2}(k-1)\log(\alpha k/(n+\alpha k)) + \tfrac{1}{2}Xn/(n+\alpha k)$$

which is maximized by $\alpha = \infty$ if $X \leq k - 1$, and $\alpha = (n - 1/k)/(X - k + 1)$ if $X > k + 1$, which is the same as the estimate based on the first two moments.

Thus $\sup G(\alpha) \approx \{\tfrac{1}{2}(X - k + 1) - \tfrac{1}{2}\log[1 + (X - k + 1)/(k - 1/n)]\}^+$ which is a monotone function of $X$; its asymptotic distribution is determined from the asymptotic distribution of $X$, which is $\chi_{k-1}^2$; and Good's test statistic $G$ is just a monotone function of $X$ asymptotically, without the asymptotic behavior stated in Good (1967). Of course, a complete Bayesian notes that $P(p_i|\alpha) = (n_i + \alpha)/(n + k\alpha)$; specifies a prior density for $\alpha$; and computes $P[p_i|\mathbf{n}] = P[(n_i + \alpha)/(n + k)|\mathbf{n}]$ averaging over the posterior density of $\alpha$ given $n$. The likelihood $P(\mathbf{n}|\alpha)$ is messy enough to suggest no simple closed form expression will be available. Good (1967) uses a log-Cauchy distribution for $\alpha$.

## 10.5. Two Stage Poisson Models

Suppose $n_i$ are Poisson $\lambda_i$, and the $\lambda_i$ are drawn from some distribution $P_0$, as in 9.4. The $n_i$ are sampled from the discrete distribution with density $p_0$,

$$p_0(n) = P_0[\lambda^n e^{-\lambda}/n!].$$

The posterior mean of $\lambda_i$ given $n_i$ is $P_0[\lambda^{n_i+1} e^{-\lambda}/n_i!]/P_0[\lambda^{n_i} e^{-\lambda}/n_i!]$ which equals $(n_i + 1)p_0(n_i + 1)/p_0(n_i)$.

Thus we can compute posterior means (and variances and other moments) if we know $p_0$. If $p_0$ is not completely known, as good Bayesians, we would need a prior distribution for it; the whole data set $n_1, \ldots, n_k$ would determine a posterior distribution for $p_0$ and the estimate $P[\lambda_i|\mathbf{n}] = (n_i + 1)P[p_0(n_i + 1)/p_0(n_i)|\mathbf{n}]$.

For many Poissons $n_i$, we might estimate $p_0(n)$ by $\#[n_i = n]/\# n_i$, the maximum likelihood estimate; however this does not take advantage of smoothness induced by $p_0(n) = P_0[\lambda^n e^{-\lambda}/n!]$. See Robbins (1956).

A special case is $P_0$ gamma with density $a^\alpha \lambda^{\alpha-1} e^{-a\lambda}/\Gamma(\alpha)$. Then $p_0(n) = $

$(1 + a)^{-(\alpha+n)} \dbinom{\alpha + n - 1}{n}$, the negative binomial; and $a, \alpha$ may be estimated

from the observed $n_i$ by maximum likelihood or by the moments $Pn = a\alpha/(1 + a)$, $P(n - Pn)^2 = \alpha[a/(1 + a)]^2$. Also $P[\lambda_i|\mathbf{n}] = P[(n_i + \alpha)/(1 + a)|\mathbf{n}]$; if only we could think of a nice prior distribution $g$ of $a, \alpha$, the posterior density

$g \prod_{i=1}^k (1 + a)^{-(\alpha+n_i)} \dbinom{\alpha + n_i - 1}{n_i}$ could be used to obtain a Bayes estimate

of $\lambda$.

## 10.6. Multinomials with Clusters

In previous sections, all cells have been treated symmetrically, but it will frequently happen that some groups of probabilities $p_i$ will be expected to be similar. One possibility is that the cells are grouped in clusters $C_1, C_2, \ldots, C_j$ and then the prior density might be taken to be $\prod(\sum_{i \in C_j} p_i)^{\alpha_j - 1}$;

this is as if we had made previous observations in which $\alpha_j$ individuals occurred in the cluster $C_j$. If the clusters are hierarchical, so that $C_i$ and $C_j$ overlap only if $C_i \subset C_j$ or $C_j \subset C_i$, this model may be reformulated as a number of Dirichlet priors on conditional probabilities, and probability estimates may be simply computed. Suppose for example the multinomial is a $2 \times 3$ contingency table:

| $n_{11}$ | $n_{12}$ | $n_{13}$ |
|---|---|---|
| $n_{21}$ | $n_{22}$ | $n_{23}$ |

Let $C_1 = (11, 12, 13)$, $C_2 = (21, 22, 23)$, $C_{ij} = \{ij\}$. The prior density $\prod (\sum_{ij \in C_k} p_{ij})^{\alpha_k - 1} = (p_{11} + p_{12} + p_{13})^{\alpha_1 - 1}(p_{21} + p_{22} + p_{23})^{\alpha_2 - 1} \prod p_{ij}^{\alpha_{ij} - 1}$ may be transformed to a density on the marginal probabilities $p_{1.} = p_{11} + p_{12} + p_{13}$, $p_{2.} = 1 - p_{1.}$, and conditional probabilities $p_{i|j} = p_{ij}/p_{i.}$.

$$(p_{1.})^{\alpha_1 - 1} p_{2.}^{\alpha_2 - 1} \prod p_{ij}^{\alpha_{ij} - 1} dp_{11} dp_{12} dp_{13} dp_{21} dp_{22}$$
$$= p_1^{\alpha_1 + 1 + \Sigma(\alpha_{1j} - 1)} p_2^{\alpha_2 + 1 + \Sigma(\alpha_{2j} - 1)} \prod p_{j|i}^{\alpha_{ij} - 1} dp_{1.} dp_{1|1} dp_{2|1} dp_{1|2} dp_{2|2}$$

(Note that only five parameters appear in the differential element, since $\sum p_{ij} = 1$ .) The new density is

$$p_{1.}^{\alpha_1 + 1 + \Sigma(\alpha_{1j} - 1)} p_{2.}^{\alpha_2 + 1 + \Sigma(\alpha_{2j} - 1)} \prod p_{j|i}^{\alpha_{ij} - 1}.$$

The advantage of this formulation is that the marginal and conditional probabilities are independent, and so it is easy to do posterior computations. For example $P[p_{ij}|\mathbf{n}] = P[p_{i.}|\mathbf{n}]P[p_{i|j}|\mathbf{n}]$. See Good (1965) for other methods of generating priors for contingency tables.

## 10.7. Multinomials with Similarities

It may happen that the cells of a multinomial are ordered in such a way that neighboring probabilities are likely to be close. The prior density ensures that neighboring probabilities are not too different. Pioneering work in this area occurs in Good and Gaskins (1971, 1980) studying density functions. For multinomial probabilities, Leonard (1973) presents the following prior density.

Let $[\log p_i]$ be multivariate normal, subject to the constraint $\sum p_i = 1$; neighboring $p_i$'s are required to be highly correlated. A similar prior density is considered by Simonoff (1980), the density $\exp[A \sum_{i=1}^{k-1} \log^2 (p_i/p_{i+1})]$; the penalty function $\sum \log^2 (p_i/p_{i+1})$ ensures that $p_i$ and $p_{i+1}$ must be close.

In order to avoid the pesky dependence $\sum p_i = 1$, let us assume $n_i$ Poisson with expectation $\lambda_i$, and that the log $\lambda_i$ form a normal autoregressive process with lag one,

$$[\log \lambda_{i+1} - \mu] = \rho[\log \lambda_i - \mu] + \sigma \varepsilon_{i+1}, \quad \varepsilon_i \text{ independent } N(0, 1).$$

A simple limiting case, with $\rho = 1$, has $\log \lambda_i$ uniform, $\log \lambda_{i+1}/\lambda_i$ independent $N(0, \sigma^2)$. The posterior density with respect to Lebesgue measure on $\{\log \lambda_i\}$ is $\propto \exp[\sum n_i \log \lambda_i - \frac{1}{2}\sum \log^2 (\lambda_{i+1}/\lambda_i)/\sigma^2 - \sum \lambda_i]$. It is difficult to compute posterior means, but the posterior mode is easier to compute: the function to be maximized is called a penalized likelihood function by Good and Gaskins (1971), with penalty function $\sum \log^2 (\lambda_{i+1}/\lambda_i)$ requiring neighboring $\lambda$'s to be close. It is not feasible to estimate the $\sigma^2$ in the obvious way, to maximize the posterior density, because the inaccessible constant of proportionality includes $\sigma$.

The modal value of $u_i = \log \lambda_i$ satisfies

$$n_i + (2u_i - u_{i-1} - u_{i+1})/2\sigma^2 - e^{u_i} = 0.$$

Concavity of $\sum n_i u_i - \sum (u_{i+1} - u_i)^2/2\sigma^2 - \sum e^{u_i}$ guarantees the existence and uniqueness of a modal value. The solution may be found by a Newton–Raphson technique. Simonoff (1980) shows that for large $k$, with $n_i$ moderate, the estimates $\hat{\lambda}_i$ are weighted averages of the $n_j$ for $j$ near $i$, giving asymptotic behavior similar to kernel estimates. These techniques are related to spline fitting methods used in regression and density estimation; see for example Wahba's remarks in the discussion of Stone (1977).

An alternative prior on $\log \lambda_i$ is $\exp[-A\sum |\log(\lambda_{i+1}/\lambda_i)|]$ which specifies the absolute differences to be exponentially distributed. The model $u_i = \log \lambda_i$ maximizes $n_i u_i - A\sum |u_{i+1} - u_i| - \sum e^{u_i}$; thus $e^{u_i} = n_i - 2A$, $n_i + 2A$, $e^{u_{i+1}}$ or $e^{u_{i-1}}$. The solution may be described by a number of intervals $(I_r, J_r)$ such that $u_i$ is constant for $I_r \leqq i \leqq J_r$. If $u_{I_{r-1}} < u_{I_r} < u_{I_{r+1}}$, then $(J_r - I_r + 1)e^{u_{I_r}} = \sum_{I_r \leqq i \leqq J_r} n_i$; this is equivalent to amalgamating the cells $i$, $I_r \leqq i \leqq J_r$. Search for the optimal intervals requires techniques similar to Barlow et al. (1972). This method clusters the cells which have similar $n_i$ and is clearer in its action than the normal prior considered previously. Its asymptotic properties are unknown.

## 10.8. Contingency Tables

The entries in a contingency table may be regarded as multinomial or Poisson; the special structure of the contingency table requires special priors for the parameters. Good (1965) has many useful ideas for such priors. See also Leonard (1975).

For a two way contingency table with entries $n_{ij}$, $1 \leqq i \leqq I$, $1 \leqq j \leqq J$ and probabilities $\{p_{ij}\}$, we often expect independence $p_{ij} = p_{i.}p_{.j}$ where $p_{i.} = \sum_j p_{ij}$, $p_{.j} = \sum_i p_{ij}$. Good considers putting a prior density on the parameters $[p_{ij}/p_{i.}p_{.j}]$, which has the effect of moving all parameter estimates $\hat{p}_{ij}$ towards independence.

For large tables with ordered rows and columns, the prior density in the Poisson model, $\exp[-\sum A \log^2 [\lambda_{ij}\lambda_{i+1 j+1}/\lambda_{ij+1}\lambda_{i+1 j}]]$, with respect to

$\log \lambda_{ij}$ Lebesgue, encourages each $2 \times 2$ table of neighboring cells to be nearly independent. The modal posterior estimates of $\lambda$ are then approximate weighted averages of counts in nearby cells.

With prior density $\exp(-\sum A|\log(\lambda_{ij}\lambda_{i+1\,j+1}/\lambda_{1\,j+1}\lambda_{i+1\,j})|)$, the posterior mode requires blocks of neighboring $2 \times 2$ tables to be independent, and so breaks the contingency table into a number of (unbalanced) subtables where independence is achieved. Computations with both these techniques are formidable.

## 10.9. Problems

P1. For all $n_i$ large, find an approximate expression for the Dirichlet parameter $\alpha$ maximizing the likelihood $(\Gamma(k\alpha)/\Gamma(\alpha)^k)(\Gamma(n+1)/\Gamma(n+\alpha k))\prod_i(\Gamma(n_i+\alpha)/\Gamma(n_i+1))$.

P2. Let $\lambda_i$ be independent gamma variables with density $a(a\lambda)^{\alpha-1}\exp(-a\lambda)/\Gamma(\alpha)$. Let $n_i$ be independent Poisson with expectations $\lambda_i$. Show that $\{\lambda_i/\sum\lambda_i\}$ given $\{n_i\}$ has the same posterior distribution as $\{p_i\}$ in the multinomial model with Dirichlet prior density $\propto \prod p_i^{\alpha-1}$.

P3. In a binomial model with $n = 10$, compute the mean square error of the Bayes estimators corresponding to beta prior densities $[p(1-p)]^{\alpha-1}$ for $\alpha = -1, 0, \frac{1}{2}, 1, 10$ and sketch the risks as a function of $p$. [Hand computation will suffice.]

Obtain the distribution of $r$ (the number of successes) given $\alpha$ and estimate $\alpha$ given $r$.

If $P_0[\alpha = 1/2] = P_0[\alpha = 1] = \frac{1}{2}$, find $P[p|r]$.

E1. If $p$ can take only the values $k/N, 0 \le k \le N$, show that the proportion of successes in $n$ trials in the binomial model, is inadmissible as an estimate of $p$ with squared error loss, when $n > N$.

P4. On visiting a new cafeteria, a distinguished statistician took five cubes of sugar for his coffee. On each wrapper was pictured a bird; of the first four, the third was a cardinal but the other three were swallows; what bird is likely to appear on the fifth wrapper? (See Good (1965).)

E2. In a week books were borrowed from a library by persons in the following categories

| | |
|---|---|
| First year students | = 6 |
| Second year students | = 10 |
| Third year students | = 7 |
| Fourth year students | = 5 |
| Statistics faculty | = 3 |
| Undergraduates | = 2 |
| Other graduate students | = 8 |
| Other faculty | = 1 |
| Other persons | = 3 |

Estimate the probability that the next book is borrowed by a person in each of the above categories.

P5. For a $2 \times 2$ table, find a prior distribution on the probabilities $p_{11}, p_{12}, p_{21}, p_{22}$ so that the Bayes test for independence is Fisher's test, rejecting independence if the first observation $n_{11}$ is too large or too small given $n_{1.}, n_{.1}$.

E3. Is the estimate $\hat{p} = 0$ admissible as an estimate of $p$ in binomial problems, with mean squared error loss?

Q1. Do a two stage analysis of the multinomial model analogous to the two stage Poisson model, 10.5.

P6. In the binomial, is the maximum likelihood estimate $\hat{p}$ admissible with loss $(p - \hat{p})^2/p(1 - p)$ or $\hat{p} \log p + (1 - \hat{p}) \log(1 - p)$? Assume $0 < p < 1$.

P7. Johnson (1971). In the binomial problem, given $r$ successes in $n$ trials, admissible estimates of $p$ are of form:

$$\hat{p} = 0, \quad r \leqq L$$
$$\hat{p} = P_0[p^{r-L}(1 - p)^{U-r-1}]/P_0(p^{r-L-1}(1 - p)^{U-r-1})$$
$$\hat{p} = 1, \quad r \geqq U$$

where $-1 \leqq L < U \leqq n + 1$, and $P_0$ is not carried by $\{0, 1\}$.

P8. (Clevenson and Zidek, 1975) For $p$ independent Poissons $n_i$ with means $\lambda_i$, show that $\delta_i(\mathbf{n}) = (1 - (\beta + p - 1)/(\sum n_i + \beta + p - 1))\mathbf{n}$ beats $\mathbf{n}$ as an estimate of $\lambda$, using loss function $\sum((\delta_i - \lambda_i)^2/\lambda_i)$, for $1 \leqq \beta \leqq p - 1$.

P9. In the multinomial, show that $\{n_i/n\}$ is inadmissible for $\{p_i\}$, with squared error loss, if the parameter values satisfy $|p_i| > \varepsilon, i = 1, 2, \ldots, k$.

## 10.10. References

Abramowitz, M. and Stegun, I. A. (1964), *Handbook of Mathematical Functions.* U.S. Department of Commerce.

Barlow, R. E., Bartholomew, D. J., Bremner, J. M. and Brunk, H. D. (1972), *Statistical Inference under Order Restrictions.* New York: John Wiley.

Clevenson, M. L. and Zidek, J. V. (1975), Simultaneous estimation of the means of independent Poisson Laws, *J. Am. Stat. Ass.* **70**, 698–705.

Good, I. J. (1965), *The Estimation of Probabilities.* Cambridge, Mass: M.I.T. Press.

——(1967), A Bayesian significance test for multinomial distributions, *J. Roy. Statist. Soc.* B **29**, 399–431.

——(1975), The Bayes factor against equiprobability of a multinomial population using a symmetric Dirichlet prior, *Annals of Statistics*, **3**, 246–250.

——and Gaskins, R. (1971), Nonparametric roughness penalties for probability densities, *Biometrika* **58**, 255–277.

——(1980). Density estimation and bump hunting by the penalized likelihood method exemplified by scattering and meteorite data, *J. Am. Stat. Ass.* **75**, 42–73.

Johnson, B. M. (1971), On the admissible estimators for certain fixed sample binomial problems, *Annals of Math. Statistics* **42**, 1579–1587.

Johnson, W. E. (1932), Appendix to probability: deductive and inductive problems, *Mind* **41**, 421–423.

Leonard, T. (1973), A Bayesian method for histograms, *Biometrika* **60**, 297–308.

——(1975), Bayesian estimation methods for two-way contingency tables, *J. Roy. Stat. Soc. B* **37**, 23–37.

Perks, W. (1947), Some observations on inverse probability including a new indifference rule, *J. Inst. Actuaries* **73**, 285–312.

Robbins, H. E. (1956), An empirical Bayes approach to statistics, *Proc. III Berkeley Symposium*, 157–163.

Simonoff, J. S. (1980), A penalty function approach to smoothing large sparse contingency tables, Ph.D. Thesis, Yale University.

Stone, C. J. (1977), Consistent non-parametric regression, *Annals of Statistics* **5**, 595–645.

# CHAPTER 11

# Asymptotic Normality of Posterior Distributions

## 11.0. Introduction

Suppose $X_1, \ldots, X_n$ are independent observations from $P_\theta$, $\theta \in R$. Suppose that $P_\theta$ has density $f_\theta(x)$ with respect to some measure $\nu$. The maximum likelihood estimate of $\theta$ (or the value of $\theta$ that maximizes the density of the posterior probability relative to the prior probability), maximizing $\prod_{i=1}^{n} f_\theta(X_i)$ is denoted by $\hat{\theta}_n$. As $n \to \infty$, Fisher established that $\hat{\theta}_n$ is asymptotically normal with mean $\theta_0$ and variance $(nI(\theta_0))^{-1}$, where $\theta_0$ is the true value of $\theta$, and $I(\theta_0)$ is Fisher's information—$\{-(d^2/d\theta^2)P_{\theta_0}[\log f_\theta(X)]\}_{\theta=\theta_0}$. The asymptotic normality requires a tedious list of regularity conditions, first promulgated by Wald.

Under almost the same conditions, with the additional requirement that the prior density be positive and continuous in the neighborhood of $\theta_0$, the posterior distribution of $\theta$ given $X_1, \ldots, X_n$ is asymptotically normal with mean $\hat{\theta}_n$ and variance $[nI(\hat{\theta}_n)]^{-1}$.

In the same way that the posterior distribution is consistent for $\theta_0$ under very general conditions, it may be shown to be normal under very general conditions; however these elegant general conditions are often more difficult to verify than the longer maximum likelihood list.

The prior density does not affect the asymptotic distribution of $\theta$ in the terms of $O(1)$ or $O(n^{-1/2})$. It does shift the mean of the asymptotic distribution by a term $O(n^{-1})$.

Similar results hold for $k$-dimensional parameter spaces.

107

## 11.1. A Crude Demonstration of Asymptotic Normality

Let $P_0$ denote the prior probability.

The posterior distribution $P_{\mathbf{X}}$ is given by

$$P_{\mathbf{X}}Y = P_0\left[\prod f_\theta(X_i)Y(\theta)\right]/P_0\left(\prod f_\theta(X_i)\right).$$

For $\theta$ near $\hat{\theta}_n$,

$$\sum \log f_\theta(X_i) = \sum \log f_{\hat{\theta}_n}(X_i) + \tfrac{1}{2}(\theta - \hat{\theta}_n)^2 \sum \frac{d^2}{d\hat{\theta}_n^2} \log f_{\hat{\theta}_n}(X_i) + \text{small}.$$

Assume that $X_1, X_2, \ldots, X_n$ are drawn from $Q$, not necessarily a member of the family $P_\theta$, $\theta \in R$; if $Q = P_{\theta_0}$ say $\theta_0$ is the true value of $\theta$. Then

$$\sum \log f_\theta(X_i) = \log f_{\hat{\theta}_n}(X_i) + \tfrac{1}{2}(\theta - \hat{\theta}_n)^2 n \frac{d^2}{d\hat{\theta}_n^2} Q(\log f_{\hat{\theta}_n}) + \text{small}.$$

Let $P_0$ have density $p_0$ with respect to Lebesgue measure $\mu$.

Then $\log p_0(\theta) = \log p_0(\hat{\theta}_n) + \text{small}$ for $\theta$ near $\theta_n$.

And $P_{\mathbf{X}}$ has density $p_{\mathbf{X}}$ with respect to Lebesgue measure

$$p_{\mathbf{X}}(\theta) \approx \exp\left[\tfrac{1}{2}(\theta - \hat{\theta}_n)^2 n \frac{d^2}{d\hat{\theta}_n^2} Q(\log f_{\theta_n})\right] c(\mathbf{X})$$

so that $\theta$ has the asymptotic density of a normal distribution with mean $\hat{\theta}_n$ and variance $(nI(\hat{\theta}_n))^{-1}$.

It is necessary to produce regularity conditions which will validate the omission of various "small" terms.

## 11.2. Regularity Conditions for Asymptotic Normality

See also Walker (1969).

**Theorem.**

(i) *Observations $X_1, X_2, \ldots, X_n$ are drawn from the unitary probability $Q$ on $\mathcal{Y}$.*

(ii) *It is contemplated that $X_1, \ldots, X_n$ might be drawn from $P_\theta$, some $\theta \in R$. It is assumed that $P_\theta$ has density $f_\theta(X)$ with respect to a measure $v$ on $\mathcal{Y}$.*

(iii) *The function $Q(\log f_\theta)$ has a unique maximum at $\theta = \theta_0$. (If $Q = P_{\theta_0'}$, necessarily $\theta_0 = \theta_0'$.)*

(iv) *The prior $P_0$ has a density $p_0$ with $p_0(\theta_0) > 0$, $p_0$ continuous at $\theta_0$.*

(v) *$P_x[|\theta - \theta_0| > \varepsilon] \to 0$ as $Q$ each $\varepsilon > 0$.*

(vi) *In a neighborhood of $\theta_0$, the derivatives $(d/d\theta) \log f_\theta, (d^2/d\theta^2) \log f_\theta$ exist and are continuous in $\theta$, uniformly in $X$.*

(vii) *$Q[(d/d\theta_0) \log f_{\theta_0}]^2 < \infty, Q((d^2/d\theta_0^2) \log f_{\theta_0}) < 0.$*

(viii) *Let $\hat{\theta}_n$ be the maximum likelihood estimate for $\theta$ near $\theta_0$ (necessarily unique as $n \to \infty$). Let $\phi_n = (\theta - \hat{\theta}_n)[-\sum(d^2/d\theta^2) \log f_\theta(X_i)]^{1/2}_{\theta = \hat{\theta}_n}$. Then the posterior density of $\phi_n$ with respect to Lebesgue measure satisfies*

$$\sup_{|\phi_n| \leq K} \left| \frac{p_n(\phi_n)}{\frac{1}{\sqrt{2\pi}} \exp(-\frac{1}{2}\phi_n^2)} - 1 \right| \to 0 \quad \text{as } Q \text{ for each } K > 0.$$

PROOF. (1) First it will be shown that $\sum \log f_\theta(X_i)$ is maximized by a unique $\hat{\theta}_n$ in a small neighborhood of $\theta_0$, as $Q$ as $n \to \infty$, with $(d/d\hat{\theta}_n)\sum \log f_{\hat{\theta}_n}(X_i) = 0$.

Since $(d/d\theta) \log f_\theta$ is continuous in $\theta$ uniformly in $X$, $(d/d\theta)Q(\log f_\theta) = Q((d/d\theta) \log f_\theta)$. Since $Q(\log f_\theta)$ has a unique maximum at $\theta = \theta_0$,

$$(d/d\theta_0)Q(\log f_{\theta_0}) = 0.$$

Let $Q[(d^2/d\theta_0^2) \log f_{\theta_0}(X_i)] = -\Delta_0$.

$$\frac{1}{n}\sum \frac{d^2}{d\theta^2} \log f_\theta(X_i) = \frac{1}{n}\sum \frac{d^2}{d\theta_0^2} \log f_{\theta_0}(X_i) + \Delta,$$

where $\Delta < \Delta_0/2$ whenever $|\theta - \theta_0| < \delta$, by uniform continuity, (vi). Since

$$\frac{1}{n}\sum \frac{d^2}{d\theta_0^2} \log f_{\theta_0}(X_i) \to Q\left[\frac{d^2}{d\theta_0^2} \log f_{\theta_0}(X)\right] \quad \text{as } Q,$$

$$\frac{1}{n}\sum \frac{d^2}{d\theta^2} \log f_\theta(X_i) < -\Delta_0/2$$

whenever $|\theta - \theta_0| < \delta$, for all large $n$ as $Q$.

Thus $\sum \log f_\theta(X_i)$ has at most one maximizing value in $|\theta - \theta_0| < \delta$ as $Q$. Also

$$\frac{1}{n}\sum \frac{d}{d\theta} \log f_\theta(X_i) = \frac{1}{n}\sum \frac{d}{d\theta_0} \log f_{\theta_0}(X_i) + \frac{1}{n}(\theta - \theta_0)\sum \frac{d^2}{d\theta^{*2}} \log f_{\theta^*}(X_i)$$

where $|\theta - \theta^*|, |\theta_0 - \theta^*| \leq |\theta_0 - \theta|$. As $n \to \infty$, $(1/n)\sum(d/d\theta_0) \log f_{\theta_0}(X_i) \to 0$ as $Q$.

Thus

$$\frac{1}{n}\sum \frac{d}{d\theta} \log f_\theta(X_i) < (\theta - \theta_0)\Delta/2 \quad \text{for } |\theta - \theta_0| < \delta, \theta > \theta_0 \text{ as } Q$$

$$\frac{1}{n}\sum \frac{d}{d\theta} \log f_\theta(X_i) > (\theta - \theta_0)\Delta/2 \quad \text{for } |\theta - \theta_0| < \delta, \theta < \theta_0 \text{ as } Q.$$

Thus $\sum(d/d\theta) \log f_\theta(X_i)$ has a zero in $|\theta - \theta_0| < \delta$ as $Q$ as $n \to \infty$, and the zero is unique because $\sum(d^2/d\theta^2)\log f_\theta(X_i) < -\Delta_0/2$ in $|\theta - \theta_0| < \delta$ as $Q$; the zero at $\theta = \hat\theta_n$ will maximize $\sum \log f_\theta(X_i)$; since $\hat\theta_n$ lies in $|\theta - \theta_0| < \delta$ for $n$ large enough, for each choice of $\delta$, $\hat\theta_n \to \theta_0$ as $Q$.

(2) $\hat\theta_n$ is asymptotically normal with mean $\theta_0$ and variance $\sigma^2/n$ where $\sigma^{-2} = -Q[(d^2/d\theta_0^2)\log f_{\theta_0}(X)]$.

$$0 = \frac{d}{d\hat\theta_n}\sum \log f_{\hat\theta_n}(X_i) = \frac{d}{d\theta_0}\sum \log f_{\theta_0}(X_i) + (\hat\theta_n - \theta_0)\frac{d^2}{d\theta_0^2}\sum \log f_{\theta_0}(X_i)$$

$$+ n\varepsilon(\theta, \mathbf{X})$$

where $\varepsilon(\theta, \mathbf{X}) \to 0$ as $Q$ from (vi).

Now $(1/n)(d/d\theta_0)\sum \log f_{\theta_0}(X)$ is asymptotically normal

$$N[0, (1/n)Q((d/d\theta_0)\log f_{\theta_0})^2]$$

and

$$(1/n)(d^2/d\theta_0^2)\sum \log f_{\theta_0}(X_i) \to Q[(d^2/d\theta_0^2)\log f_{\theta_0}(X)] \quad \text{as } Q.$$

Thus $\hat\theta_n - \theta_0$ is asymptotically normal with mean 0 and variance $\sigma^2/n$.

(3) To conclude, let $\delta_n \to 0$;

$$\log p_{\mathbf{X}}(\theta)/p_{\mathbf{X}}(\hat\theta_n) = \log p_0(\theta)/p_0(\hat\theta_n) + \sum \log f_\theta(X_i) - \sum \log f_{\hat\theta_n}(X_i)$$

$$= \varepsilon(\theta, \hat\theta_n) + \tfrac{1}{2}(\theta - \hat\theta_n)^2\left(\sum \frac{d^2}{d\hat\theta_n^2}\log f_{\hat\theta_n}(X_i) + n\varepsilon(\theta, \mathbf{X})\right)$$

where by (iv) and (vi), $\varepsilon(\theta, \hat\theta_n) \to 0$ as $Q$ and $\varepsilon(\theta, \mathbf{X}) \to 0$ as $Q$, uniformly over $|\theta - \theta_0| \le \delta_n$.

From (vii), $(1/n)\sum(d^2/d\hat\theta_n^2)\log f_{\hat\theta_n}(X_i) \to -\Delta$, and $\phi_n$ is a linear transformation of $\theta$, with $\phi_n(\hat\theta_n) = 0$. Thus $p_{\mathbf{X}}(\theta)/p_{\mathbf{X}}(\hat\theta_n) = p_n(\phi_n)/p_n(0)$,

(A) $\log p_n(\phi_n)/p_n(0) + \tfrac{1}{2}\phi_n^2 \to 0$ as $Q$ uniformly over $|\phi_n| \le K$.

(B) Also $\log p_n(\phi_n)/p_n(0) < -\tfrac{1}{4}(\theta - \hat\theta_n)^2 n\Delta$ for all $|\theta - \theta_n| \le \delta_n$ as $Q$ as $n \to \infty$.

(C) Finally, from (v) $P_{\mathbf{X}}(|\theta - \theta_0| > \delta_n) \to 0$ as $Q$ for some $\delta_n \to 0$.

It is necessary to combine facts (A), (B), (C) to determine $p_n(0)$. From (C),

$$\int_{|\theta-\theta_0|\le\delta_n} p_n(\phi_n)d\phi_n \to 1$$

From (B),

$$\int_{|\theta-\theta_0|\le\delta_n, |\theta-\hat\theta_n|>K/\sqrt{n}} p_n(\phi_n)d\phi_n \le p_n(0)\int_{|\theta-\hat\theta_n|>K/\sqrt{n}} \exp[-\tfrac{1}{4}(\theta-\hat\theta_n)^2 n\Delta]d\phi_n$$

$$\le p_n(0)C_n\exp[-\tfrac{1}{4}K^2\Delta] \quad \text{for large } n.$$

From (A),

$$p_n^{-1}(0)\int_{|\theta-\hat\theta_n|<K/\sqrt{n}} p_n(\phi_n)d\phi_n \to \lim \int_{|\theta-\hat\theta_n|<K/\sqrt{n}} \exp(-\tfrac{1}{2}\phi_n^2)d\phi_n$$

$$= (\sqrt{2\pi} - C_n'\exp[-\tfrac{1}{2}K^2\Delta]).$$

Here $C_n$ and $C'_n$ are bounded by 1 as $K \to \infty$, $n \to \infty$, so $p_n(0)\sqrt{2\pi} \to 1$.

Thus $p_n(\phi_n)/[\exp(-\frac{1}{2}\phi_n^2)/\sqrt{2\pi}] - 1 \to 0$ uniformly over $|\phi_n| \leq K$ as required.                                                                    $\square$

*Notes*: The conditions of the theorem look forbidding but they are merely those conditions which permit neglect of "small" terms in the Taylor series expansion. It is not necessary that the $X_1, \ldots, X_n$ be sampled from a member of the family $P_\theta$, but it is necessary that a unique member $P_{\theta_0}$ be "closest" to $Q$ in maximizing $Q[\log dP_\theta/dQ]$; if there are $\theta_0, \theta'_0$ such that $P_{\theta_0}$ and $P_{\theta'_0}$ are both closest, then the limiting posterior distribution should be bimodal with modes near $\theta_0$ and $\theta'_0$. The regularity conditions on $\log f_\theta$ near $\theta_0$ are far stronger than is necessary. It is necessary that the prior density be positive at $\theta_0$, and that it be continuous at $\theta_0$ (the conclusion of the theorem requires $p_X(\theta_n)/p_X(\theta_0) \to 1$ as $\sqrt{n}(\theta_n - \theta_0) \to 0$, $n \to \infty$; continuity of $(d/d\theta) \log f_\theta(X)$ requires $p_0(\theta_n)/p_0(\theta_0) \to 1$ as $\sqrt{n}(\theta_n - \theta_0) \to 0$, which requires continuity of $p_0$).

It is necessary that the posterior distribution concentrates at $\theta_0$; maximum likelihood conditions for convergence of $\hat{\theta}_n$ to $\theta_0$ might be given, governing the behavior of the likelihood outside neighborhoods of $\theta_0$, but it may be easier to check convergence of the posterior distribution directly.

## 11.3. Pointwise Asymptotic Normality

**Theorem.**

(i) Let $X_1, X_2, \ldots, X_n$ be sampled from a unitary $Q$ on $\mathcal{Y}$.

(ii) Let $P_\theta$, $\theta \in R$ be a family of probabilities on $\mathcal{Y}$ with densities $f_\theta$ with respect to some measure $v$ on $\mathcal{Y}$.

(iii) Let $\theta = \theta_0$ be a local maximum of $Q(\log f_\theta)$.

(iv) Let a prior probability $P_0$ have density $p_0$ positive and continuous at $\theta_0$.

(v) $(d/d\theta_0) \log f_{\theta_0}$ exists as $Q$ and $Q((1/\delta) \log(f_{\theta_0+\delta}/f_{\theta_0}) - (d/d\theta_0) \log f_{\theta_0})^2 \to 0$ as $\delta \to 0$.

(vi) $Q((d/d\theta_0) \log f_{\theta_0})^2 < \infty$;     $(d^2/d\theta_0^2)Q(\log f_{\theta_0}) = -1/v < 0$.

*Then the posterior density $p_X$ of $\theta$ is pointwise asymptotically normal in the neighborhood of $\theta_0$ with mean $\theta_0 + (v/n)\sum(d/d\theta_0) \log f_{\theta_0}(X_i)$ and variance $v/n$; that is*

$$p_X(\theta_0 + \xi\sqrt{v/n})/p_X(\theta_0) - \phi\left[\xi - \sqrt{\frac{v}{n}}\sum\frac{d}{d\theta_0}\log f_{\theta_0}(X_i)\right] \Bigg/$$

$$\phi\left[\sqrt{\frac{v}{n}}\sum\frac{d}{d\theta_0}\log f_{\theta_0}(X_i)\right] \to 0$$

*in Q-probability for each $\xi$, where $\phi(u) = \exp(-\frac{1}{2}u^2)/\sqrt{2\pi}$.*

PROOF. Let

$$h_\delta(X) = \frac{1}{\delta} \log f_{\theta_0 + \delta}(X) / f_{\theta_0}(X)$$

$$h_0(X) = \frac{d}{d\theta_0} \log f_{\theta_0}(X) = \lim_{\delta \to 0} h_\delta(X) \quad \text{defined as } Q.$$

Let $\qquad \delta_n = \xi \sqrt{v/n}.$

Then $p_X(\theta_0 + \delta_n)/p_X(\theta_0) = [p_0(\theta_0 + \delta_n)/p_0(\theta_0)] \times \prod [f_{\theta_0 + \delta_n}(X_i)/f_{\theta_0}(X_i)]$

$\qquad L_n(\xi) = \log p_X(\theta_0 + \delta_n)/p_X(\theta_0) = \log p_0(\theta_0 + \delta_n)/p_0(\theta_0) + \sum \delta_n h_{\delta_n}(X_i)$

From (iv), $L_n(\xi) - \sum \delta_n h_{\delta_n}(X_i) \to 0$ as $n \to \infty$.

Also, $\sum \delta_n [h_{\delta_n}(X_i) - h_0(X_i)]$ has mean $n\delta_n Q(h_{\delta_n} - h_0)$ and variance $n\delta_n^2 Q(h_{\delta_n} - h_0)^2 \to 0$ as $n \to \infty$ by (v).

From (iii) and (v), $(d/d\theta_0)Q(\log f_{\theta_0}) = Q((d/d\theta_0) \log f_{\theta_0}) = Qh_0 = 0$.

From (vi), $Q(\log f_{\theta_0 + \delta_n}) = Q(\log f_{\theta_0}) + \delta_n(d/d\theta_0)Q(\log f_{\theta_0}) + \frac{1}{2}\delta_n^2[- 1/v + \varepsilon\delta_n]$ where $\varepsilon\delta_n \to 0$ as $n \to \infty$. Thus $n\delta_n Q(h_{\delta_n} - h_0) + \frac{1}{2}n\delta_n^2/v \to 0$ as $n \to \infty$.

Therefore $L_n(\xi) - \delta_n \sum h_0(X_i) + (1/2)n\delta_n^2/v \to 0$ in $Q$-probability.

$$(*) \qquad L_n(\xi) - \xi \sqrt{\frac{v}{n}} \sum h_0(X_i) + \frac{1}{2}\xi^2 \to 0 \quad \text{in } Q\text{-probability}.$$

$$\log\{p_X(\theta_0 + \xi \sqrt{v/n})/p_X(\theta_0)\} - \log\left\{ \phi\left[ \xi - \sqrt{\frac{v}{n}} \sum h_0(X_i) \right] \right/ $$
$$\phi\left[ \sqrt{\frac{v}{n}} \sum h_0(X_i) \right] \right\} \to 0$$

in $Q$-probability.

$$p_X(\theta_0 + \xi\sqrt{v/n})/p_X(\theta_0) - \phi\left[ \xi - \sqrt{\frac{v}{n}} \sum h_0(X_i) \right] \bigg/ \phi\left[ \sqrt{\frac{v}{n}} \sum h_0(X_i) \right] \to 0$$

in $Q$-probability as required.                    □

*Notes*: The condition (iii) is weaker than the corresponding condition (iii) of Theorem 11.2; also there is no condition corresponding to 11.2(v) which requires the posterior distribution to concentrate on $\theta_0$. Thus the posterior density may be asymptotically normal in the neighborhood of $\theta_0$ without concentrating there!

Condition 11.3(v) is much weaker than 11.2(vi); thus it is only possible to prove pointwise convergence of the posterior density, rather than uniform convergence. The core of the proof is showing that the log posterior density is parabolic near the "optimal" $\theta_0$, as in equation (*).

Note that the expression for asymptotic variance is $(d^2/d\theta_0^2)Q(\log f_{\theta_0})$ rather than $Q((d^2/d\theta_0^2) \log f_{\theta_0})$; the second derivatives of $\log f_\theta$ may not exist for many $X$ and $\theta$, but the second derivatives of $Q(\log f_\theta)$, averaging out $X$, may well exist. See Ibragimov and Khas'minskii (1973) and LeCam (1970) for some related results in maximum likelihood asymptotics.

## 11.4. Asymptotic Normality of Martingale Sequences

**Theorem.** *Let* $\mathcal{X}_0 \subset \mathcal{X}_1 \subset \cdots \subset \mathcal{X}_n \subset \cdots \subset \mathcal{X}$ *be probability spaces, and let* $P_0$ *be a unitary probability on* $\mathcal{X}$. *Let* $P_i$ *be probabilities on* $\mathcal{X}$ *to* $\mathcal{X}_i$ *with* $P_i = P_i P_j$, $i \leq j$. *Let X be an element of* $\mathcal{X}_\infty$, *where* $\mathcal{X}_\infty$ *is the minimal complete probability space with a probability* $P_\infty$ *equal to* $P_0$ *on* $\mathcal{X}_n$ *each n. Define*

$$u_n = P_n X - P_{n-1} X$$
$$s_n^2 = P_n(X - P_n X)^2$$

*Assume* (i) $0 < s_n^2 < \infty$ *all n, as* $P_0$.

(ii) $\sum_{j=n+1}^{\infty} P_{j-1}(u_j^2)/s_n^2 \to 1$ *as* $n \to \infty$, *as* $P_0$.

(iii) $\sup_{j>n} P_{j-1}(u_j^2)/s_n^2 \to 0$ *as* $n \to \infty$, *as* $P_0$.

(iv) $\sum_{j=n+1}^{\infty} P_{j-1}|u_j|^3/s_n^3 \to 0$ *as* $n \to \infty$, *as* $P_0$.

*Then*

$$P_n f[(X - P_n X)/s_n] \to \int \frac{1}{\sqrt{2\pi}} f(u) e^{-(1/2)u^2} du, \quad \text{as } P_0.$$

*for each bounded continuous function f (so that X is asymptotically normal with mean* $P_n X$ *and variance* $s_n^2$ *given* $\mathcal{X}_n$).

PROOF. The proof parallels the usual method for proving a central limit theorem for sums. Here $X - P_n X = \sum_{j=n+1}^{\infty} u_j = \lim_{N \to \infty} \sum_{j=n+1}^{N} u_j$; the quantities $\{u_j\}$ play the role of the summands in the central limit theorem; they are not independent, but satisfy $u_j$ $P_0$-uncorrelated with $f(u_k)$ for $k > j$, $f$ measurable.

Let $f_n(u) = \exp(itu/s_n)$.

Then

$$P_n[f_n(u_{n+1} + u_{n+2})/P_n f_n(u_{n+1}) P_{n+1} f_n(u_{n+2})] = P_n P_{n+1}[f_n(u_{n+1})/P_n f_n(u_{n+1})]$$
$$= 1$$

By induction $P_n[f_n(\sum_{i=n+1}^{N} u_i)/\prod_{i=n+1}^{N} P_{i-1} f_n(u_i)] = 1$ (this makes the characteristic function of $P_N X - P_n X$ nearly the same as a product). From Theorem 4.2, $P_n|P_N X - X| \to 0$ as $N \to \infty$.

Also $|f_n(x) - f_n(y)| \leq |x - y|$, and $|f_n(x)| \leq 1$, $|f_n^{-1}(x)| \leq 1$. Therefore $|P_n([f_n(\sum_{n+1}^{N} u_i) - f_n(X - P_n X)]/\prod_{i=n+1}^{N} P_{i-1} f_n(u_i))| \leq P_n|P_N X - X| \to 0$ as $N \to \infty$.

$$P_{i-1} f_n(u_i) = 1 + \frac{it}{s_n} P_{i-1} u_i - \tfrac{1}{2} t^2 P_{i-1} u_i^2/s_n^2 + v|t|^3 P_{i-1}|u_i|^3/s_n^3 \quad \text{with } |v| \leq 1$$

$$\sum_{i=n+1}^{N} \log P_{i-1} f_n(u_i) = -\tfrac{1}{2} t^2 \sum_{i=n+1}^{N} P_{i-1}(u_i^2)/s_n^2 + v|t|^3 \sum_{n+1}^{N} P_{i-1}|u_i|^3/s_n^3 + \xi_n$$

where $\xi_n \to 0$ as $n \to \infty$ by (iii).

[Using the facts $x - \frac{1}{2}x^2 < \log(1 + x) < x$,

$$\sum x_i - \sup|x_i|\sum|x_i| < \sum \log(1 + x_i) < \sum x_i.]$$

Thus $\sum_{i=n+1}^{\infty} \log P_{i-1} f_n(u_i)$ exists by (ii) and (iv), and approaches $-\frac{1}{2}t^2$ as $n \to \infty$ as $P_0$. Since $\sum_{i=N+1}^{\infty} \log P_{i-1} f_n(u_i) \to 0$ as $N \to \infty$,

$$P_n\left[ f_n(X - P_n X)\bigg/ \prod_{i=n+1}^{\infty} P_{i-1} f_n(u_i) \right]$$

$$= P_n\left[ f_n(X - P_n X)\bigg/ \prod_{i=n+1}^{N} P_{i-1} f_n(u_i) \times \prod_{N+1}^{\infty} P_{i-1} f_n(u_i) \right]$$

$$= P_n\left[ f_n(X - P_n X)\bigg/ \prod_{i=n+1}^{N} P_{i-1} f_n(u_i) Z_N \right] \quad \text{where } Z_N \to 1 \text{ as } N \to \infty$$

$$= \lim_{N \to \infty} P_n\left[ f_n(X - P_n X)\bigg/ \prod_{i=n+1}^{N} P_{i-1} f_n(u_i) \right] = 1.$$

$$P_n[f_n[X - P_n X]] \to e^{-(1/2)t^2} \quad \text{as } n \to \infty, \text{ since} \quad \prod_{i=n+1}^{\infty} P_{i-1} f_n(u_i) \to e^{-(1/2)t^2}.$$

Thus the characteristic function of $(X - P_n X)/s_n$ approaches the characteristic function of the unit normal as $Q$. Since any continuous function that is zero outside a compact set can be uniformly approximated on the set by a finite sum $\sum a_j \exp(it_j u)$, the same result holds for such continuous functions. Extension to arbitrary bounded continuous functions is straightforward. □

*Notes*: These results are very free of regularity conditions, and especially of independence conditions. The conditions on the increments in posterior means $P_n X - P_{n-1} X$ might be difficult to verify. Conditions (iv) might be weakened to the Lindeberg-like condition $\sum_{j=n+1}^{\infty} P_{j-1}[|u_j|^2 \{|u_j| > \varepsilon\}]/s_n^2 \to 0$. See Hall and Heyde (1980), and Brown (1971).

EXAMPLE. Suppose $\mathscr{X}_n$ is generated by $X_1, \ldots, X_n$, a sample from the Bernoulli distribution $P[x] = p^{\{x=1\}}(1 - p)^{\{x=0\}}$ given $p$, and where $p$ has some prior distribution $P_0$. See Awad (1978), p. 53. Letting $r = \sum X_i$,

$$P_n(p) = f(r, n) = P_0[p^{r+1}(1 - p)^{n-r}]/P_0[p^r(1 - p)^{n-r}]$$
$$u_{n+1} = f(r + X_{n+1}, n + 1) - f(r, n)$$

$u_{n+1}$ given $\mathscr{X}_n$ has values $f(r + 1, n + 1) - f(r, n)$ with probability $f(r, n)$,
$$f(r, n + 1) - f(r, n) \text{ with probability } 1 - f(r, n).$$

$P_n(u_{n+1}) = 0$, so $f(r, n + 1)[1 - f(r, n)] + f(r + 1, n + 1)f(r, n) = f(r, n)$.

$$s_n^2 = P_n(p - P_n p)^2 = f(r + 1, n + 1)f(r, n) - f(r, n)^2$$
$$P_n(u_{n+1}^2) = s_n^4/f(r, n)(1 - f(r, n))$$
$$P_n|u_n|^3 = s_n^6/[f^{-2}(r, n) + (1 - f(r, n))^{-2}].$$

Note that $f(r, n) = P_n p \to p$ as $n \to \infty$, as $P_0$.

*Assume that $ns_n^2 \to p(1 - p)$, as $P_0$.*

Then $\displaystyle\sum_{n+1}^{\infty} P_{j-1}(u_j^2)/s_n^2 - n \sum_{n+1}^{\infty} [p(1 - p)/(j - 1)^2]/p(1 - p) \to 0.$

Since

$$n \sum_{j=n+1}^{\infty} 1/(j - 1)^2 \to 1 \quad \text{as } n \to \infty, \text{ (ii) is satisfied.}$$

$$\sup_{j>n} P_{j-1}(u_j^2)/s^2 \le (1 + \varepsilon)/n \to 0 \quad \text{as } n \to \infty, \text{ satisfying (iii)}$$

$$\sum_{j=n+1}^{\infty} P_{j-1}|u_j|^3/s_n^3 \to \frac{n^{3/2}}{[p(1 - p)]^{3/2}} \sum_{j=n+1}^{\infty} \frac{[p(1 - p)]^3}{(j - 1)^3} \Big/ \left[\frac{1}{p^2} + \frac{1}{(1 - p)^2}\right] \to 0$$

satisfying (iv).

Thus the posterior distribution of $p$ given $x_1, \ldots, x_n$ is normal whenever $ns_n^2 \to p(1 - p)$, as $P_0$, that is, whenever the posterior variance converges to the asymptotic variance of the maximum likelihood estimator.

## 11.5. Higher Order Approximations to Posterior Densities

Let's just blast away with Taylor series expansions and leave the regularity conditions till later. See Johnson (1967) and Hartigan (1965).

(i) *Assume $X_1, \ldots, X_n$ are a sample from $P_\theta$ having density $f_\theta$ with respect to some measure $v$ on $\mathcal{Y}$. Let $\theta$ be a real valued random variable.*

Let $\quad h_r(X) = [d/d\theta]^r \log f_\theta(X)_{\theta = \theta_0}.$

$\quad g_r = Q[h_r(X)]$ with respect to some measure $Q$ on $\mathcal{Y}$.

(ii) *Assume $Q(\log f_\theta)$ is maximal at $\theta = \theta_0$.*
(iii) *Assume the prior $P_0$ on $S$ has density $p_0$, and the posterior $P_X$ has density $p_X$.*
Then

$$\log p_X(\theta) = \log p_X(\theta_0) + \log p_0(\theta)/p_0(\theta_0) + \sum_{i=1}^{n} \log f_\theta(X_i)/f_{\theta_0}(X_i)$$

$$\log [p_X(\theta)/p_X(\theta_0)] = (\theta - \theta_0)\frac{d}{d\theta_0} \log p_0(\theta_0)$$

$$+ (\theta - \theta_0)\sum h_1(X_i) + \tfrac{1}{2}(\theta - \theta_0)^2 \sum h_2(X_i)$$

$$+ \tfrac{1}{6}(\theta - \theta_0)^3 \sum h_3(X_i)$$

$$+ o[(\theta - \theta_0)] + o[n(\theta - \theta_0)^3]$$

(iv) This expansion is justified by requiring the first three derivatives $[d/d\theta]^r \log f_\theta(X)$ to be continuous in a neighborhood of $\theta_0$, uniformly in $X$; and by requiring the derivatives $(d/d\theta) \log p(\theta)$ to be continuous in a neighborhood of $\theta_0$. In order to ensure that large deviations $|\theta - \theta_0|$ have negligible probability, assume $P_X[|\theta - \theta_0| > n^{-1/2+\varepsilon}]n^k \to 0$ for every $k > 0$. The later terms are negligible if $Qh_2 < 0$.

$$P_X(\theta) = c(X) \exp\left\{\tfrac{1}{2}\sum h_2(X_i)\left[\theta - \theta_0 + \left[\sum h_1(X_i) + \frac{d}{d\theta_0}\log p_0\right]\middle/\sum h_2(X_i)\right]^2\right\}$$
$$\times \left\{1 + \tfrac{1}{6}(\theta - \theta_0)^3 h_3(X_i) + o(\theta - \theta_0) + o[n(\theta - \theta_0)^3]\right\}.$$

Here the term $\tfrac{1}{6}(\theta - \theta_0)^3 \sum h_3(X_i)$ causes an $O(n^{-1/2})$ skewness departure from normality. The only effect of the prior is in shifting the mean by $-(d/d\theta_0)\log p_0/\sum h_2(X_i)$. The first three moments determine the asymptotic distribution

$$P_n\theta = \theta_0 - \sum h_1(X_i)/\sum h_2(X_i)$$
$$- \left\{\frac{d}{d\theta_0}\log p_{\theta_0} + \tfrac{1}{2}\left[(\sum h_1)^2 - \sum h_2\right]\sum h_3/(\sum h_2)^2\right\}\middle/\sum h_2 + O(n^{-2})$$
$$P_n(\theta - P_n\theta)^2 = (\sum h_1 \sum h_3/\sum h_2 - \sum h_2)^{-1} + O(n^{-2})$$
$$P_n(\theta - P_n\theta)^3 = -\sum h_3/(\sum h_2)^3 + O(n^{-3})$$

## 11.6. Problems

E1. Show that the binomial model satisfies conditions 11.2, when the prior density is continuous and positive at $p_0$, $0 < p_0 < 1$, and $p_0$ is assumed true.

E2. In the binomial case, if the prior distribution has an atom at $p_0$, show that the conditions of 11.4 are not satisfied.

E3. Let $f[X_1, X_2, \ldots, X_n]$ be the marginal density of the observations, and let $p(\theta)$ be the prior density. Show, under conditions 11.2, that

$$f(X_1, X_2, \ldots, X_n)/\prod f(X_i|\hat{\theta}_n) \to p(\theta_0)\sqrt{2\pi}\left\{nQ\left[-\frac{\partial^2}{\partial\theta_0^2}\log f\right]\right\}^{-1/2}$$

P1. Observations $X$ are $N(\mu, 1)$ and $\mu$ has prior density uniform on $|\mu| \geq 1$. Give the asymptotic behavior of the posterior distribution as the true $\mu_0$ ranges from $-\infty$ to $\infty$.

E4. Under the conditions 11.2, when $\theta_0$ is true, show that $P(\theta < \theta_0|X_n)$ is asymptotically uniformly distributed.

E5. Under the conditions 11.2, the posterior distribution of $\log[\prod f(X_i|\theta)\prod f(X_i|\hat{\theta}_n)]$ is asymptotically $-\tfrac{1}{2}\chi_1^2$. [Bayes intervals for $\theta$ thus coincide approximately with maximum likelihood intervals.]

E6. $X_1, \ldots, X_n$ are uniform over $[\theta - \tfrac{1}{2}, \theta + \tfrac{1}{2}]$ and $\theta$ is uniform over $-\infty$ to $\infty$. Give the asymptotic behavior of the posterior density when $\theta = 0$.

E7. $f(x|\theta) = 1/\theta \qquad 0 < x < \theta$
$\qquad = 1/(1-\theta) \quad$ if $\theta \leqq x < 1$
$\qquad = 0 \qquad\qquad$ elsewhere.

If $\theta$ is uniform over $(0, 1)$, specify the asymptotic behavior of the posterior density of $\theta$ when $\theta = 1/2$ is true.

P2. $f(x|\mu) = \frac{1}{2} \exp\{-|x-\mu|\}$, $\mu$ uniform. Specify the asymptotic behavior of the posterior density of $\mu$, given $\mu = 0$.

P3. Let $g$ be such that $g(X)$ and $g^2(X)$ are $P_\theta$ integrable. If $X_1, \ldots, X_n$ is a sample from $P_\theta$, and $P$ on $\theta$ is unitary, show that

$$P(n \text{ var } [P_\theta[g(X)]|\mathbf{X}]) \leqq PP_\theta[g^2(X)] - P[P_\theta^2 g(X)].$$

Thus $P_\theta g(X)$ is known, given $\mathbf{X}$, to order $n^{-1/2}$.

E8. Generalise theorem 11.2 to $k$-dimensional parameters.

P4. For the binomial model, the prior on $p$ has density $\frac{2}{3}\{0 \leqq p \leqq \frac{1}{2}\} + \frac{4}{3}\{\frac{1}{2} \leqq p \leqq 1\}$. What is the posterior distribution of $p$ asymptotically, when the true value is $p = \frac{1}{2}$?

P5. The observation $X, Y$ is bivariate normal, means $a(\theta), b(\theta)$, identity covariance matrix, $a(\theta) = 0$ for $\theta \leqq 0$, $a(\theta) = \theta$ for $\theta \geqq 0$, and $b(\theta) = a(-\theta)$. Find the asymptotic posterior distribution of $\theta$, when the true value of the means is $(1, 1)$. Assume a uniform prior distribution for $\theta$.

P6. In the binomial model, find nondegenerate prior distributions for $p$ for which $n \operatorname{var}(p|\mathbf{X}_n) \to 0$.

P7. For $X_1, \ldots, X_n$ from $N(\mu, \sigma^2)$, prior $\mu \sim N(\mu_0, \sigma_0^2)$ verify the conditions of the martingale central limit theorem.

Q1. Let $X_1, \ldots, X_n$ be from the normal mixture $pN(\mu_1, \sigma_0^2) + (1-p)N(\mu_2, \sigma_0^2)$ where $p$ has uniform prior, $\mu_1$ and $\mu_2$ are independently $N(0, 1)$, $\sigma_0^2$ fixed. What is the asymptotic posterior distribution of $p, \mu_1, \mu_2$ for various true values of $p, \mu_1$ and $\mu_2$?

Q2. Let $\mathscr{X}_1 \subset \mathscr{X}_2 \subset \ldots \subset \mathscr{X}_n \ldots$ be increasing, $\theta \in \mathscr{X}_\infty$, and suppose $Z_n \in \mathscr{X}_n$, $Z_n \to \theta$ has the property that $(Z_n - \theta)/\sigma_n(\theta) \to N(0, 1)$ in distribution given $\theta$. Show that $(\theta - Z_n)/\sigma_n(Z_n) \to N(0, 1)$ in distribution given $Z_n$, provided $\theta$ has continuous positive density on the line. (Note: $Z_n$ may not have a convergent density.) [Here $\sigma_n(\theta)$ is the standard deviation of $\theta$ given $\mathscr{X}_n$ and $\sigma_n(Z_n)$ is the standard deviation of $Z_n$ given $\mathscr{X}_n$.]

P8. Let $X_0, X_1, X_2, \ldots$ be observations from an autoregressive process $X_t = \alpha X_{t-1} + \xi_t$ where the $\xi_t$ are i.i.d. normal. Assume $\alpha$ is uniform on $(-1, 1)$. Find the asymptotic behavior of the posterior distribution of $\alpha$ given $X_0, X_1, X_2, \ldots, X_n$.

P9. Let $X_1, \ldots, X_n$ be a sample from the density $\exp(\theta - X)$, $X \geqq \theta$. Let $\theta$ have a prior density which is continuous and positive at $\theta = 0$. Find the asymptotic distribution of $\theta$ given $X_1, \ldots, X_n$ if $X_1, \ldots, X_n$ are sampled from the uniform on $(0, 1)$.

## 11.7. References

Awad, A. M. (1978), *A martingale approach to the asymptotic normality of posterior distributions*, Ph.D. Thesis, Yale University.

Brown, B. M. (1971), The martingale central limit theorem, *Ann. Math. Statist.* **42**, 59–66.

Hall, P. and Heyde, C. C. (1980), *Martingale Limit Theory and Its Applications.* New York: Academic Press.

Hartigan, J. A. (1965), The asymptotically unbiased prior distribution, *Ann. Math. Statist.* **36**, 1137–1154.

Ibragimov, I. A. and Khas'minskii, R. Z. (1973), Asymptotic behaviour of some statistical estimators II Limit theorems for the a posteriori density and Bayes estimators, *Theor. Probability Appl.* **18**, 76–91.

——(1975), Local asymptotic normality for non-identically distributed observations, *Theor, Probability Appl.* **20**, 246–260.

Johnson, R. A. (1967), An asymptotic expansion for posterior distributions, *Ann. Math. Statist.* **38**, 1899–1907.

LeCam, L. (1958), Les proprietés asymptotiques des solutions des Bayes, *Publ. Inst. Statist. Univ. Paris* **7**, 17–35.

——(1970), On the assumptions used to prove asymptotic normality of maximum likelihood estimates, *Ann. Math. Statist.* **41**, 802–828.

Walker, A. M. (1969), Asymptotic behavior of posterior distributions, *J. Roy. Stat. Soc.* B **31**, 80–88.

# CHAPTER 12

# Robustness of Bayes Methods

## 12.0. Introduction

A statistical procedure is robust if its behavior is not very sensitive to the assumptions which justify it. In classical statistics these are assumptions about a probability model $\{P_\theta, \theta \in \Theta\}$ for the observations in $\mathcal{Y}$, and about a loss function $L$ connecting the decision and unknown parameter value. In Bayesian statistics, there is in addition an assumed prior distribution.

Bayesian techniques have been used by Box and Tiao (1973) and others to study classical robustness questions such as the choice of a good estimate of a location parameter for "near-normal" distributions; they imbed the normal in a family with one more parameter, and then use standard Bayesian techniques to determine the posterior distribution of the location parameter. The usual robustness studies allow for a much larger neighborhood of distributions however.

In studying Bayesian robustness, we wish to evaluate the effect on the posterior distribution and on Bayesian decisions of various components of the probability model. Since the loss function is chosen by the decision maker it seems plausible to concentrate on the probability parts of the model.

(i) the likelihood component $\{P_\theta, \theta \in \Theta\}$
(ii) the prior component $P_0$

Here we consider mainly the prior component $P_0$, using the techniques of de Robertis and Hartigan (1981).

## 12.1. Intervals of Probabilities

Let $Q_1$ and $Q_2$ be probabilities on $\mathcal{X}$.

Define $Q_1 \leqq Q_2$ if $Q_1 X \leqq Q_2 X$ whenever $X \geqq 0$.

The *interval* of probabilities $(L, U)$ is the set of probabilities $Q$ with $L \leqq Q \leqq U$. The probability $L$ will be called the *lower* probability, and the probability $U$ will be called the *upper* probability. If $L$ has density $l$, and $U$ has density $u$ with respect to $v$, then $(L, U)$ consists of the probabilities with density $q$, $l \leqq q \leqq u$.

**Theorem.** Inf $\{Q(Y)/Q(X) | L \leqq Q \leqq U\}$ *is the unique solution* $\lambda$ *of* $U(Y - \lambda X)^- + L(Y - \lambda X)^+ = 0$, *provided* $UX^- + LX^+ > 0$.

PROOF. Note that $X^+ = X\{X(s) \geqq 0\}$, $X^- = X\{X(s) \leqq 0\}$.

Since $Q(Y) \geqq UY^- + LY^+$ for $L \leqq Q \leqq U$, and $Q_0 Z = U[\{Y \leqq 0\}Z] + L[\{Y \geqq 0\}Z]$ satisfies $L \leqq Q_0 \leqq U$, $\inf\{Q(Y) | L \leqq Q \leqq U\} = Q_0 Y = UY^- + LY^+$.

Now $\inf Q(Y)/Q(X) \geqq \lambda$ if and only if $\inf[Q(Y) - \lambda Q(X)] \geqq 0$, since $Q(X) \geqq 0$ for $L \leqq Q \leqq U$.

Thus $\inf Q(Y)/Q(X) \geqq \lambda$ if and only if $U(Y - \lambda X)^- + L(Y - \lambda X)^+ \geqq 0$.

Also

$$\frac{d}{d\lambda}\{U(Y - \lambda X)^- + L(Y - \lambda X)^+\}$$

$$= U(-X\{Y \leqq \lambda X\}) + L[-X\{Y \geqq \lambda X\}].$$

$$\leqq -UX^- - LX^+ < 0$$

(since $Q_0 Z = U\{Y \leqq \lambda X\}Z + L\{Y \geqq \lambda X\}Z$ lies in $(L, U)$).

Thus $U(Y - \lambda X)^- + L(Y - \lambda X)^+$ is strictly decreasing, and is zero at $\lambda = \inf Q(Y)/Q(X)$ as required.                                    $\square$

## 12.2. Intervals of Means

**Theorem.** *Let* $L = N(0, 1)$, $U = kL$.

*For* $Q \in (L, U)$, *the mean of* $Q$ *is* $QX/Q1$ *where* $X(s) = s$.

*Then* $QX/Q1$ *has range* $[-\gamma(k), \gamma(k)]$ *where* $\gamma(k)$ *satisfies*

$$k\gamma = (k - 1)[\phi(\gamma) + \gamma\Phi(\gamma)]$$

*where* $\phi(x) = \exp(-\tfrac{1}{2}x^2)/\sqrt{2\pi}$, $\Phi(x) = \int_{-\infty}^x \phi(u)du$.

PROOF. From 12.1, $\inf QX/Q1$ is the solution of $U(X - \lambda)^- + L(X - \lambda)^+ = 0$. That is

$$\int_{x \leqq \lambda} (x - \lambda)k\phi(x)dx + \int_{x \geqq \lambda} (x - \lambda)\phi(x)dx = 0$$

that is

$$- k\phi(\lambda) - k\lambda\Phi(\lambda) + \phi(\lambda) - \lambda[1 - \Phi(\lambda)] = 0$$
$$\lambda = (k - 1)[\phi(\lambda) + \lambda\Phi(\lambda)]$$
$$- k\lambda = (k - 1)[\phi(-\lambda) - \lambda\Phi(-\lambda)]$$

Thus $- \gamma(k) = \inf QX/Q1$. Similarly $\gamma(k) = \sup QX/Q1$. $\qquad\square$

| $k$ | 1 | 1.25 | 1.50 | 1.75 | 2 | 2.5 | 3 | 4 | 5 | 6 | 7 | 8 | 9 | 10 |
|---|---|---|---|---|---|---|---|---|---|---|---|---|---|---|
| $\gamma(k)$ | 0 | .089 | .162 | .223 | .276 | .364 | .436 | .549 | .636 | .707 | .766 | .817 | .862 | .901 |

Thus quite substantial changes in the probability Q do not affect the mean too much. Similar Bayes estimates will arise from a wide range of priors.

## 12.3. Intervals of Risk

**Theorem.** *Suppose that the risk $r(d, \theta)$ is the loss in making decision d when $\theta$ is true. Let the Bayes risk $B(Q) = \inf_d (Q[r(d, \theta)]/Q(1))$, the probable loss when the best decision is taken. Assume $0 < L(1) \leq U(1) < \infty$.*
  $\text{Inf}\{B(Q)|L \leq Q \leq U\}$ *is the unique solution of*

$$\beta_1(\lambda) = \inf_d [U(r(d, \theta) - \lambda)^- + L(r(d, \theta) - \lambda)^+] = 0$$

$\text{Sup}\{B(Q)|L \leq Q \leq U\}$ *is no greater than the unique solution of*

$$\beta_2(\lambda) = \inf_d [L(r(d, \theta) - \lambda)^- + U(r(d, \theta) - \lambda)^+] = 0.$$

PROOF. It is straightforward to show that $\beta_1(\lambda)$ and $\beta_2(\lambda)$ are continuous, strictly decreasing and have unique zeroes $\lambda_1$ and $\lambda_2$

$$\inf\{B(Q)|L \leq Q \leq U\} = \sup\{\lambda|B(Q) \geq \lambda \quad \text{for Q such that } L \leq Q \leq U\}$$

$$= \sup\{\lambda|Q[r(d, \theta)] \geq \lambda Q(1)$$
$$\text{for all } d \text{ and } Q, L \leq Q \leq U\}$$
$$= \sup\{\lambda|Q[r(d, \theta) - \lambda] \geq 0$$
$$\text{all } d \text{ and } Q, L \leq Q \leq U\}$$
$$= \sup\{\lambda|\inf_d(L[r(d, \theta) - \lambda]^+ + U[r(d, \theta) - \lambda]^-) \geq 0\}$$
$$= \sup\{\lambda|\beta_1(\lambda) \geq 0\} = \lambda_1$$

$$\sup\{B(Q)|L \leq Q \leq U\} = \inf\{\lambda|B(Q) < \lambda \quad \text{for } Q, L \leq Q \leq U\}$$
$$= \inf\{\lambda|Q[r(d, \theta)] < \lambda Q(1) \quad \text{some } d,$$
$$\text{for each } Q, L \leq Q \leq U\}$$
$$\leq \inf\{\lambda|\sup_Q Q[r(d, \theta) - \lambda] < 0 \quad \text{some } d\}$$

$$= \inf\{\lambda \mid L[r(d, \theta) - \lambda]^- + U[r(d, \theta) - \lambda]^+ < 0$$
$$\text{some } d\}$$
$$= \inf\{\lambda \mid \beta_2(\lambda) < 0\} = \lambda_2. \qquad\qquad \square$$

## 12.4. Posterior Variances

For the normal location problem, take $L = N(0, 1)$, $U = kL$.
For $r(d, \theta) = (d - \theta)^2$, $B(Q)$ is the variance of $\theta$.
It is bounded by the solutions of

$$\inf_d U\{k[(d - \theta)^2 - \lambda_1]^- + [(d - \theta)^2 - \lambda_1]^+\} = 0$$
$$\inf_d U\{k[(d - \theta)^2 - \lambda_2]^- + [(d - \theta)^2 - \lambda_2]^-\} = 0.$$

By symmetry of $U$ about 0, the optimal solution is $d = 0$ for both equations.
(Consider probabilities of form $k\phi\{(d - \theta)^2 < \lambda_1\} + \phi\{(d - \theta)^2 > \lambda_1\}$; note
that the mean value of such probabilities lies between $d$ and 0 for $k > 1$;
this implies that the only solution to the first equation is $d = 0$. For the
second equation, $k[(d - \theta)^2 - \lambda_2]^+ + [(d - \theta)^2 - \lambda_2]^+$ is convex so its
minimizing value is unique. By symmetry if $d$ is a minimum, so is $- d$.
Therefore $d = 0$.)

The solutions $\lambda_1$ and $\lambda_2$ are posterior variances of elements of $(L, U)$ so
the bounds of Theorem 12.3 are sharp.

$$\lambda_1 = \Delta_1^2 \text{ is the solution of } \Delta\phi(\Delta) + (\Delta^2 - 1)\left[\Phi(\Delta) - \frac{k - 1}{2k - 2}\right] = 0$$

$$\lambda_2 = \Delta_2^2 \text{ is the solution of } \Delta\phi(\Delta) + (\Delta^2 - 1)\left[\Phi(\Delta) - \frac{2k - 1}{2k - 2}\right] = 0$$

| $k$ | | 1.25 | 1.50 | 1.75 | 2 | 2.5 | 3 | 4 | 5 | 6 | 7 | 8 | 9 | 10 |
|---|---|---|---|---|---|---|---|---|---|---|---|---|---|---|
| $\Delta_1$ | 1 | .947 | .904 | .870 | .840 | .792 | .754 | .697 | .654 | .621 | .574 | .592 | .552 | .535 |
| $\Delta_2$ | 1 | 1.055 | 1.100 | 1.140 | 1.174 | 1.233 | 1.282 | 1.360 | 1.421 | 1.472 | 1.515 | 1.552 | 1.585 | 1.615 |

Thus, again, a very large change in the probability density causes a relatively
minor change in posterior variance. [Factor of 10 for density gives factor of
2 for variance.]

## 12.5. Intervals of Posterior Probabilities

**Lemma.** $L \leq cQ \leq U$ for some $c$, if and only if $QX/QY \leq UX/LY$ for each
$X \geq 0$, $Y \geq 0$.

PROOF. The "only if" is obvious.

If $QXLY \leq UXQY$    all $X, Y \geq 0$

$$\sup_{X \geq 0} Q(X)/UX = c_1 \leq c_2 = \inf_{Y \geq 0} QY/LY$$

$$c_2 LX \leq QX \leq c_1 UX$$

$$c_2 LX \leq QX \leq c_2 UX$$

$$L \leq \frac{1}{c_2} Q \leq U \quad \text{as required.}$$

**Theorem.** *Let* $X, Y$ *be random variables satisfying the conditions of Bayes'* *theorem* (3.4):

(i) $X, Y$ *and* $X \times Y$ *are random variables from* $U, \mathscr{L}$ *to* $S, \mathscr{X}, T, \mathscr{Y}$ *and* $S \times T, \mathscr{X} \times \mathscr{Y}$.
(ii) $f$ *is a density on* $\mathscr{X} \times \mathscr{Y}$.
(iii) $\mathscr{X}$ *and* $\mathscr{Y}$ *are* $\sigma$-*finite.*
(iv) $f_T(t): s \to f(s, t) \in \mathscr{X}$ *each* $t$.
(v) $P_{XS}^Y g = R^Y(g f_S)$ *for some probability* $R$ *on* $\mathscr{Y}$.
(vi) *For each* $Q^X, L^X \leq Q^X \leq U^X, f/Q^X f_T$ *is a density on* $\mathscr{X} \times \mathscr{Y}$.

*Then the quotient probability* $Q_Y^X$ *corresponding to the prior probability* $Q^X$ *satisfies, for some* $k(Y)$,

$$L_Y^X \leq Q_Y^X k(Y) \leq U_Y^X (U^X f_T / L^X f_T).$$

PROOF. By Bayes' theorem,

$$Q_Y^X h = Q^X(h f_T)/Q^X(f_T)$$

$$Q_Y^X h_1 / Q_Y^X h_2 = Q^X(h_1 f_T)/Q^X(h_2 f_T) \leq U^X(h_1 f_T)/L^X(h_2 f_T).$$

From the lemma,

$$(L^X f_T) L_Y^X \leq k(Y) Q_Y^X \leq (U^X f_T) U_Y^X. \qquad \square$$

# 12.6. Asymptotic Behavior of Posterior Intervals

**Theorem.**

(i) *Let* $X, Y_1, Y_2, \ldots, Y_n, \ldots$ *be random variables from* $\mathscr{L}$ *to* $\mathscr{X}, \mathscr{Y}_1, \ldots,$ $\mathscr{Y}_n, \ldots,$ *and assume that* $Y_i^{-1}(\mathscr{Y}_i)$ *is increasing.*
(ii) *Let the quotient probability* $P_X^{Y_n}[g] = R^{Y_n}(g f_S^n)$ *for some* $f^n$ *which is a* *density with respect to* $\mathscr{X} \times \mathscr{Y}_n$ *on* $S \times T_n$, *and some probability* $R$ *on* $\mathscr{Y}_n$. *Assume that these quotient probabilities agree with a conditional* *probability* $P_X$ *defined on the smallest probability space including all* $Y_i^{-1}(\mathscr{Y}_i): P_X g_n(Y_n) = P_X^{Y_n} g_n$ *for* $g_n \in \mathscr{Y}_n$.
(iii) *Assume that* $P_X^{Y_n}$ *is unitary.*

(iv) *Let $L^X$, $U^X$ be unitary probabilities such that $f^n/L^X(f^n_{T_n})$ is a density with respect to $\mathscr{X} \times \mathscr{Y}_n$, and $L^X g = U^X[lg]$ where $U^X\{l = 0\} = 0$.*

(v) *Assume that $g_0 \in \mathscr{X}$ and $l \in \mathscr{X}$ are $\mathscr{Y}_n$-approximable in $U$-probability that is, $k_n \in \mathscr{Y}_n$, $k'_n \in \mathscr{Y}'_n$,*

$$U^X P^{Y_n}_X |g_0 - k_n| \to 0,$$
$$U^X P^{Y_n}_X |l - k'_n| \to 0.$$

*Then* $\displaystyle\sup_{L^X \leqq Q^X \leqq U^X} |Q^X_{Y_n} g_0 - g_0(X)| \to 0$ *almost surely.*

PROOF. Assume $l(X) > 0$ without loss of generality. Let $A$ be the set of values of $X$ such that, for all rational $\lambda$,

$$U^X_{Y_n}\{l[g_0 - \lambda]^+ + [g_0 - \lambda]^-\} \to \{l(X)[g_0 - \lambda]^+ + [g_0 - \lambda]^-\}$$

as $P_X$. By Doob's Theorem (Doob, 1949), $U^X(A^c) = 0$.

For a fixed value of $X$ in $A$, suppose $g_0(X) > \alpha$, rational. Then $l(X)[g_0(X) - \alpha]^+ + [g_0(X) - \alpha]^- > 0$, so $U^X_{Y_n}\{l[g_0 - \alpha]^+ + [g_0 - \lambda]^-\} > 0$ for large $n$, so $\displaystyle\sup_{L \leqq Q \leqq U} Q^X_{Y_n} g_0 > \alpha$ for all large $n$, as $P_X$. Similarly, if $g_0(X) < \beta$ for $\beta$ rational, $\displaystyle\inf_{L \leqq Q \leqq U} Q^X_{Y_n} g_0 < \beta$ all large $n$, as $P_X$. Since these results hold for all rational $\alpha$ and $\beta$, $\displaystyle\sup_{L \leqq Q \leqq U} |Q^X_{Y_n}(g_0(X)) - g_0| \to 0$ as $P_X$, except for a set of $X$ values of $U$ probability zero. □

*Note:* A similar theorem is proved in deRobertis and Hartigan (1981) with $L$, $U$ $\sigma$-finite.

## 12.7. Asymptotic Intervals under Asymptotic Normality

**Theorem.** *Let $X$, $Y_1, \ldots, Y_n, \ldots$ be random variables satisfying the conditions of Theorem 12.6, and, in addition, assume that $g_0(X)$ is asymptotically conditionally normal under $U$:*

$$U^X_{Y_n} c[(g_0 - \mu_n)/\sigma_n] \to \int c(u) \exp\left(-\frac{1}{2}u^2\right) du/\sqrt{2\pi} \quad as \ U,$$

*for each bounded continuous $c$, where $\mu_n = U^X_{Y_n} g_0$, $\sigma^2_n = U^X_{Y_n} g^2_0 - \mu^2_n$. Then*

$$\sigma_n^{-1}\left[\sup_{L \leqq Q \leqq U} Q^X_{Y_n} g_0 - (\mu_n + \sigma_n \gamma(k_n))\right] \to 0 \quad as \ U,$$

$$\sigma_n^{-1}\left[\inf_{L \leqq Q \leqq U} Q^X_{Y_n} g_0 - (\mu_n - \sigma_n \gamma(k_n))\right] \to 0 \quad as \ U,$$

*where $\gamma(k)$ is the solution of $k\gamma = (k-1)[\phi(\gamma) + \gamma\Phi(\gamma)]$, $k_n = 1/U^X_{Y_n} l(X)$.*

PROOF. By 12.6, $k_n = 1/U_{Y_n}^X l \to 1/l(X)$ as $U$.

By asymptotic normality of $g_0(X)$, for each $\lambda$,

$$U_{Y_n}^X \left\{ \left[ \frac{g_0 - \mu_n}{\sigma_n} - \lambda \right]^+ + l \left[ \frac{g_0 - \mu_n}{\sigma_n} - \lambda \right]^- \right\}$$

$$\to \int \left\{ [u - \lambda]^+ + l[u - \lambda]^- \right\} \phi(u) du \quad \text{as } U.$$

Let $A$ be the set $X$ values for which $l(X) > 0$, and the above convergence occurs for all rational $\lambda$. If $\gamma[1/l(X)] < \alpha$, then $\int \{(u - \alpha)^+ + l(X)(u - \alpha)^-\} \phi du < 0$, so that $U_{Y_n}^X \{ [(g_0 - \mu_n)/\sigma_n - \alpha]^+ + l[(g_0 - \mu_n)/\sigma_n - \alpha]^- \} < 0$ all large $n$, so $\sup Q_{Y_n}^X [(g_0 - \mu_n)/\sigma_n] > \alpha$ all large $n$. If $\gamma[1/l(X)] > \alpha$, then $\sup Q_{Y_n}^X [(g_0 - \mu_n)/\sigma_n] > \alpha$ all large $n$. Also $l(k_n) \to \gamma[1/l(X)]$ as $U$. Thus

$$\sigma_n^{-1} \sup [Q_{Y_n}^X g_0 - \mu_n - \sigma_n \gamma(k_n)] \to 0 \quad [U]. \qquad \square$$

*Note*: This theorem permits a close approximation to the interval of posterior means $Q_{Y_n}^X g_0(X)$, computed by assuming that $g_0(X)$ has upper probability $N(\mu_n, \sigma_n^2)$ and lower probability $N(\mu_n, \sigma_n^2) U_{Y_n}^X [l(X)]$.

## 12.8. A More General Range of Probabilities

If $L \leqq Q \leqq U$, and the measures have densities $l$, $q$, $u$, then $l \leqq q \leqq u$ and $q(s_1)/q(s_2) \leqq u(s_1)/l(s_2)$. This formulation has the advantage of permitting $Q$ to be a unitary probability. A difficulty in the present interval of probabilities is that $q$ may be dramatically discontinuous.

More generally let $q(s_1)/q(s_2) \leqq u(s_1, s_2)$ define $Q \in R$.

The function $u(s_1, s_2)$ might be such that $u(s_1, s_2) \to 1$ as $s_1 \to s_2$.

Necessarily $u(s_1, s_2) \leqq u(s_1, s_3) u(s_3, s_2)$.

The posterior density $q_t(s_1)/q_t(s_2) \leqq (f_t(s_1)/f_t(s_2)) u(s_1, s_2)$, so posterior densities are handled in the same framework. It is sometimes convenient to use $\log[q(s_1)/q(s_2)] \leqq \rho(s_1, s_2)$. Then $\rho(s_1, s_2) \leqq \rho(s_1, s_3) + \rho(s_2, s_3)$. (Note that $\rho$ may be negative, so it is not a metric.)

For conditional densities $f_s(t)$ it seems desirable to constrain movements in $f_{s_1}(t)$ and $f_{s_2}(t)$ where $s_1$ and $s_2$ are close. This suggests the baroque $(f_{s_1}(t_1)/f_{s_2}(t_1)) \cdot (f_{s_2}(t_2)/f_{s_1}(t_2)) \leqq u(s_1, s_2, t_1, t_2)$. Again posterior densities obey a bound of the same type. Maybe $u(s_1, s_2, t_1, t_2) = U(s_1, s_2) V(t_1, t_2)$ would be viable, but it doesn't force $u(s_1, s_2, t_1, t_2) = 1$ if $s_1 = s_2$ or $t_1 = t_2$.

It is necessary to decide if $QX \geqq 0$ all $Q \in R$. Let $A^c = \{s | X(s) \geqq 0\}$, $A = \{s | X(s) < 0\}$. Then $q(s) = \inf_{s' \in A^c} q(s') u(s, s')$ for $s \in A$; and $q(s') = \sup_{s \in A} q(s)/u(s, s')$ for $s' \in A^c$, in a solution which minimizes $QX$. But I can't see any simple way to characterize $X$ with $QX \geqq 0$, and such a characterization is really necessary to use the range.

## 12.9. Problems

E1. A prior distribution for the binomial parameter $p$ is such that no interval of length $I$ has more than twice the probability of any other interval of length $I$, for all $I$. Show $Pp \geq \sqrt{2} - 1$.

E2. Let $X_1, \ldots, X_n$ be $n$ observations from $N(\theta, 1)$. Let the prior for $\theta$ be $Q$, $L \leq Q \leq U$ where $L$ is Lebesgue measure and $U = 2L$. Show that the posterior mean lies in the interval $\bar{X} \pm .276/\sqrt{n}$.

P1. Suppose 7 successes are observed in 10 binomial trials. Let the prior for $p$ lie between $U =$ uniform $(0, 1)$ and $2U$. Find the posterior mean's range. [Hint use the binomial cumulative distribution.]

P2. A prior distribution for the binomial parameter $p$ lies between $U(0, 1)$ and $2U(0, 1)$. Find the range of the variance of $p$.

P3. Consider densities of form

$$f_d = c(d)\frac{1}{\sqrt{2\pi}}\exp(-\tfrac{1}{2}s^2)[k\{|d - s|^2 \leq \lambda\} + \{|d - s|^2 \geq \lambda\}], \qquad k \geq 1.$$

Find the value of $d$ for which the density $f_d$ has minimum variance.

P4. Suppose $r$ successes are observed in $n$ binomial trials. Let the prior for $p$ lie between $U(0, 1)$ and $2U(0, 1)$. Find an asymptotic expression for the interval of posterior means.

P5. Let $f = 1/\sqrt{2\pi}\exp[-\tfrac{1}{2}x^2 + \varepsilon(x)]$ where $|\varepsilon(x)| \leq 1$. Find bounds for the posterior density of $\theta$ given $X_1, \ldots, X_n$ where $X_1, \ldots, X_n$ is a sample from $f(x - \theta)$, and $\theta$ has uniform prior density.

## 12.10. References

Box, G. E. P. and Tiao, G. C. (1973), *Bayesian Inference in Statistical Analysis*. Reading: Addison–Wesley.

Doob, J. L. (1949), Applications of the theory of martingales, *Colloques Internationaux de Centre National de la Recherche Scientific Paris* 22–28.

DeRobertis, L. and J. A. Hartigan (1981), Bayessian inference using intervals of measures, *The Annals of Statistics* **9**, 235–244.

# CHAPTER 13

# Nonparametric Bayes Procedures

## 13.0. Introduction

Whereas Bayes procedures require detailed probability models for observations and parameters, nonparametric procedures work with a minimum of probabilistic assumptions. It is therefore of interest to examine nonparametric problems from a Bayesian point of view.

Usually nonparametric procedures apply to samples of observations from an unknown distribution function $F$. Inferences are made which are true for all continuous $F$. For example if $x_{(1)}, x_{(2)}, \ldots, x_{(n)}$ denote order statistics of the sample, $[x_{(k)}, x_{(k+1)}]$ is a confidence interval for the population median of size $\binom{n}{k} 2^{-n}$, provided the true $F$ is continuous.

We must give some sort of family of distributions over distribution functions $F$ which can be used as priors and posteriors in a Bayesian approach. Ferguson (1973) suggests the Dirichlet process, which for a general observation space $\mathcal{Y}$, gives a distribution over probabilities $P$ on $\mathcal{Y}$ such that $P(B_1), \ldots, P(B_k)$ is Dirichlet whenever $\{B_i\}$ is a partition of the sample space.

No unitary prior is known to reproduce nonparametric confidence procedures; worse, no prior of any sort is known that reproduces such confidence procedures. However some confidence procedures correspond to families of conditional probabilities, and Lane and Sudderth (1978) have used finitely additive probabilities to generate confidence procedures.

## 13.1. The Dirichlet Process

Let $\mathcal{Y}$ on $T$ be a probability space, let $\mathcal{P}$ be the set of unitary probabilities $P$ on $\mathcal{Y}$, and let $\mathcal{X}$ denote the smallest probability space on $\mathcal{P}$ such that $X : P \rightarrow PY$ lies in $\mathcal{X}$ for all $Y$ in $\mathcal{Y}$. A Dirichlet process $D_\alpha$ on $\mathcal{X}$, corresponding to a bounded measure $\alpha$ on $T(\alpha(T) < \infty)$, is such that $PB_1, PB_2, \ldots, PB_k$ is distributed as a Dirichlet $D_{\alpha(B_1),\alpha(B_2),\ldots,\alpha(B_k)}$ for each partition $B_1, B_2, \ldots, B_k$ of $T$. Proofs that a Dirichlet process exists are given in Ferguson (1973) and Blackwell and MacQueen (1973).

Following Blackwell and MacQueen, a sequence of random variables $Y_1, Y_2, Y_3, \ldots$ taking values in $\mathcal{Y}$, $T$ is a *Pólya sequence with parameter* $\alpha$ if

$$P[f(Y_i)] = \alpha(f)/\alpha(T), \quad \text{for } f \in \mathcal{Y}$$
$$P_i f = P[f(Y_{i+1}) \mid Y_1, Y_2, \ldots, Y_i] = [\alpha(f) + \sum_{j \leq i} f(Y_j)]/[\alpha(T) + i].$$

Given $Y_1, Y_2, \ldots, Y_i$, the distribution of $Y_{i+1}$ is a mixture, in the proportion of $i$ to $\alpha(T)$, of the empirical distribution based on $Y_1, Y_2, \ldots, Y_i$ and the distribution $\alpha/\alpha(T)$. As $i$ approaches $\infty$, the empirical component predominates and the limiting distribution of $Y_{i+1}$ given $Y_1, Y_2, \ldots, Y_i$ is the limiting frequency distribution of $Y_1, Y_2, \ldots, Y_i$. This limiting distribution, when it exists, will be taken to be a realization $P$ of the Dirichlet process. The different limiting distributions $P$, for different sequences $Y_1, Y_2, \ldots, Y_n, \ldots$ give a distribution of probabilities $P$ that satisfy the definition of the Dirichlet process.

**Theorem** (Blackwell–MacQueen). *Let $\mathcal{Y}$ on $T$ be separable (there exist a sequence of $0$–$1$ functions $A_1, A_2, \ldots, A_n, \ldots$ such that $\mathcal{Y}$ is the smallest probability space including $A_1, A_2, \ldots, A_n, \ldots$). Let $\{Y_i\}$ be a Pólya sequence with parameter $\alpha$. For each $Y_1, Y_2, \ldots, Y_i, \ldots$, define*

$$P^*f = \lim_{n \to \infty} \frac{\sum_i f(Y_i)}{n} \quad \text{when the limit exists for all } f \text{ in } \mathcal{Y}$$

$$P^*f = \alpha f/\alpha T \quad \text{when the limit does not exist for all } f \text{ in } \mathcal{Y}.$$

*Then $P^*$ is distributed as a Dirichlet process $D_\alpha$ on $\mathcal{X}$, and the conditional distribution of $Y_1, Y_2, \ldots, Y_i, \ldots$ given $P^*$ is such that the $Y_i$ are independent each with distribution $P^*$.*

PROOF. If $Y_1, Y_2, Y_3, \ldots$ is a Pólya sequence, the Ionescu Tulcea theorem (Neveu, 1965, p. 162) states that a probability $P$ exists on the product space $T \times T \ldots, \mathcal{Y} \times \mathcal{Y} \ldots$ such that $P$ is consistent with each of the conditional probabilities $P_i$. The separability of $\mathcal{Y}$ guarantees that functions of the form $\{Y_1 = Y_2\}$ lie in $\mathcal{Y} \times \mathcal{Y}$ provided $\mathcal{Y}$ includes all singleton functions $\{t\}, t \in T$. $[\{Y_1 \neq Y_2\} = \sup_i |\{Y_1 \in A_i\} - \{Y_2 \in A_i\}| \in \mathcal{Y} \times \mathcal{Y}$; otherwise there exist $y_1 \neq y_2$ such that $y_1, y_2$ lie both inside or both outside of every $A_i$, and $\mathcal{Y}$ generated by $A_i$ consists of functions $f$ with $f(y_1) = f(y_2)$; thus the singleton function $\{y_1\}$ is excluded from $\mathcal{Y}$.] If $\mathcal{Y}$ does not include all singletons, consider the space $\mathcal{Y}^*$, $T^*$ where $T^*$ consists of the equivalence classes $B_t, t \in T$; $t' \in B_t$ if and only if $\{t \in A_i\} = \{t' \in A_i\}$ all $A_i$. And $\mathcal{Y}^*$ consists of the functions $f^*(B_t) = f(t), f \in \mathcal{Y}$. Note that $f^*$ is well defined since $f(t) = f(t')$

whenever $t' \in B_t$. Now $\mathscr{Y}^*$ is separable and includes all singleton functions and the theorem may be proved for $\mathscr{Y}^*$. The Dirichlet process $P^*$ defined on $\mathscr{Y}^* \times \mathscr{Y}^* \times \ldots$ has the desired properties. It will therefore be assumed that $\mathscr{Y}$ includes singletons.

It will be shown first that $P^*f = \lim_{n \to \infty} \sum f(Y_i)/n$ except for sequences $\{Y_i\}$ in a set of probability zero. Let $f_1(t) = \{Y_1 = t\}$. Then $f_1(Y_n)$ is an exchangeable sequence given $Y_1$, so from de Finetti's theorem, 4.5, $\sum_{i=2}^{n} f_i(Y_i)/n$ converges to say $p_1$ with probability 1. Next let $f_2(t) = \{Y_1 = t \text{ or } Y_2 = t\}$. Then $\{f_2(Y_n), n > 2\}$ is an exchangeable sequence given $Y_1, Y_2$, and so $\sum_{i=3}^{n} f_2(Y_i)/n$ converges to say $p_2$ with probability 1. Similarly, if $f_k(t) = \bigcup_{1 \le i \le k} \{Y_i = t\}$, then $\sum f_k(Y_i)/n$ converges to say $p_k$ with probability 1, for all $k$, $1 \le k \le \infty$. Now

$$P[f_k(Y_n) \mid Y_1, Y_2, \ldots, Y_k] = [k + \alpha f_k]/[k + \alpha T] \to 1 \quad \text{as } k \to \infty$$

$$P[\sum f_k(Y_i)/(n-k) \mid Y_1, Y_2, \ldots, Y_k] = P[f_k(Y_n) \mid Y_1, \ldots, Y_k] \to 1 \quad \text{as } k \to \infty$$

$$P[\sum f_k(Y_i)/n > 1 - \varepsilon \mid Y_1, \ldots, Y_k] \to 1 \quad \text{as } k \to \infty, \text{ each } \varepsilon > 0$$

$$P[p_k > 1 - \varepsilon \mid Y_1, \ldots, Y_k] \to 1 \quad \text{as } k \to \infty.$$

A probability $P$ exists on $\mathscr{Y} \times \mathscr{Y} \times \mathscr{Y} \ldots$ consistent with these conditional probabilities, and from separability, the functions $f_k(Y_i)$ lie in $\mathscr{Y} \times \mathscr{Y} \times \mathscr{Y} \times \ldots$. Thus with probability 1, all the limits $\sum f_k(Y_i)/n$ exist and

$$\lim_{k \to \infty} \lim \sum f_k(Y_i)/n = 1.$$

This guarantees that $P^*f = \lim \sum f(Y_i)/n$ exists for all $f$ in $\mathscr{Y}$; $P^*$ is a discrete distribution carried by $\{Y_i\}$. To show this,

$$\sum_{i=1}^{n} f(Y_i)/n = f(Y_1)[\sum f_1(Y_i)/n]$$

$$+ f(Y_2)[\sum f_2(Y_i) - \sum f_1(Y_i)]/n + \cdots$$

$$+ f(Y_k)[\sum f_k(Y_i) - \sum f_{k-1}(Y_i)]/n$$

$$+ A[n - \sum f_k(Y_i)]/n \quad \text{where } |A| \le \sup |f|$$

$$\lim_{n \to \infty} |\sum f(Y_i)/n - (f(Y_1)p_1 + \cdots + f(Y_k)(p_k - p_{k-1}))| \le \sup |f|(1 - p_k).$$

Thus all the limits $\sum f(Y_i)/n$ exist if the $p_i$ exist and $\lim_k p_k = 1$. It is straightforward to show that $P^*f = \sum f(Y_i)(p_i - p_{i-1})$, $p_0 = 0$, defines a probability on $\mathscr{Y}$, for each sequence $Y_1, Y_2, \ldots, Y_i$ where the limits exist. Since $P^*f = \alpha f/\alpha T$ defines a probability $\mathscr{Y}$ when the limits don't exist, $P^*$ always takes values in $\mathscr{P}$.

To show that $P^*$ is distributed as $D_\alpha$, it is necessary to show that $P^*B_1, \ldots, P^*B_k$ is distributed as Dirichlet $D_{\alpha(B_1), \ldots, \alpha(B_k)}$ for each partition $B_1, \ldots, B_k$ of $T$.

$$P^*B_j = \lim_{n \to \infty} \sum \{Y_i \in B_j\}/n$$

Define $$Z_i = \sum j \{Y_i \in B_j\}.$$

Then $Z_i$ is a random variable taking $k$ discrete values, and it may be shown that $Z_i$ is a Pólya sequence with parameter $\alpha^*$, $\alpha^*\{i\} = \alpha(B_i)$. Since $P^*B_j = \lim \sum\{Z_i = j\}/n$, the problem is reduced to showing that $P^*$ is distributed as $D_\alpha$ when $T$ is finite. If $T = \{1, 2, \ldots, k\}$, let $P^* = \{p_1, p_2, \ldots, p_k\}$, $\sum p_i = 1$, and note from de Finetti's theorem that the $Y_i$ are independent multinomial given $P^*$, with $P[Y_i = j \mid P^*] = p_j$.

Then

$$P[p_1^{\gamma_1} p_2^{\gamma_2} \cdots p_k^{\gamma_k}] = P[\text{first } \gamma_1 \, Y's = 1,$$
$$\text{next } \gamma_2 \, Y's = 2,$$
$$\vdots$$
$$\text{last } \gamma_k \, Y's = k]$$
$$= \frac{\alpha(1)}{\alpha(T)} \cdot \frac{\alpha(1) + 1}{\alpha(T) + 1} \cdots \frac{\alpha(1) + \gamma_1 - 1}{\alpha(T) + \gamma_1 - 1} \times$$
$$\frac{\alpha(2)}{\alpha(T) + \gamma_1} \cdot \frac{\alpha(2) + 1}{\alpha(T) + \gamma_1 + 1} \cdots \frac{\alpha(2) + \gamma_2 - 1}{\alpha(T) + \gamma_1 + \gamma_2 - 1}$$

$$\cdots \cdots \cdots \cdots$$

The expression on the right is the $\gamma_1, \gamma_2, \ldots, \gamma_k^{\text{th}}$ moment of a Dirichlet distribution with parameters $\alpha(1), \alpha(2), \ldots, \alpha(k)$, $\sum \alpha(i) = \alpha(T)$. Since the Dirichlet distributions is characterized by its moments, the result follows.

It remains to be shown that the $Y_1$ are independent given $P^*$ with distribution $P^*$. It is necessary to check that $P[\prod f_i(Y_i) \mid P^*] = \prod P^*[f_i(Y_i)]$ obeys the product law: $P[\prod P^* f_i(Y_i)] = P[\prod f_i(Y_i)]$. If the $f_i$ are each members of the partition $B_1, B_2, \ldots, B_k$ this follows from de Finetti's theorem for the case $\mathcal{Y}$ finite. More general $f_i$ may be approximated by linear combinations of these simple $f_i$.                                                      $\square$

## 13.2. The Dirichlet Process on (0,1)

(1) *Let $\alpha$ be uniform.*

$F(x)$ has expectation $x$; thus $F(x)$ is beta with density $F^{x-1}(1 - F)^{-x}$.

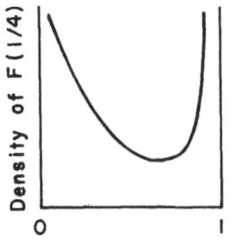

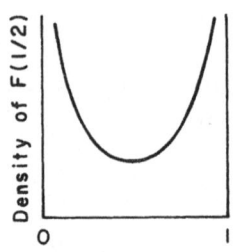

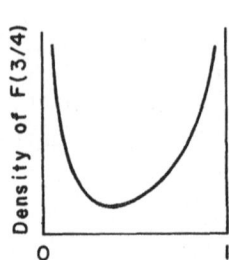

We could generate a single random $F$ from $D_\alpha$ as follows:

(a) select $F(\tfrac{1}{2})$ from $\mathrm{Be}(\tfrac{1}{2}, \tfrac{1}{2})$

(b) select $F(\tfrac{1}{4})/F(\tfrac{1}{2})$ from $\mathrm{Be}(\tfrac{1}{4}, \tfrac{1}{4})$, $[F(\tfrac{3}{4}) - F(\tfrac{1}{2})]/F(\tfrac{1}{2})$ from $\mathrm{Be}(\tfrac{1}{4}, \tfrac{1}{4})$.

(c) select $\left[F((2k+1)/2^n) - F(k/2^{n-1})\right]/\left[F((k+1)/2^{n-1}) - F(k/2^{n-1}))\right]$ from $\mathrm{Be}(1/2^n, 1/2^n)$

(d) continue forever … after you're finished:

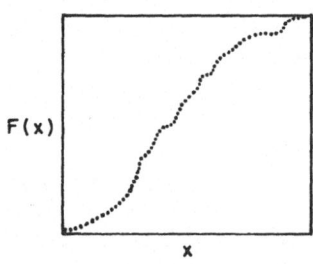

Note that $F$ will be quite bumpy, because the relative changes in $F$ will be near 0 or 1.

(2) $\alpha$ *gives weights 1 to 1/4, 1/2, 3/4, 1 and is zero elsewhere.*

$F(x)$ is $\mathrm{Be}\left[\alpha(0, x], \alpha(x, 1]\right]$. Thus $F(x) = 0$ for $x < 1/4$, and $F(x) = 1$ for $x = 1$; $F(x)$ changes value only at $x = 1/4, 1/2, 3/4, 1$ and has atoms $\Delta F(\tfrac{1}{4})$, $\Delta F(\tfrac{1}{2})$, $\Delta F(\tfrac{3}{4})$, $\Delta F(1)$ which are Dirichlet $D_{1,1,1,1}$.

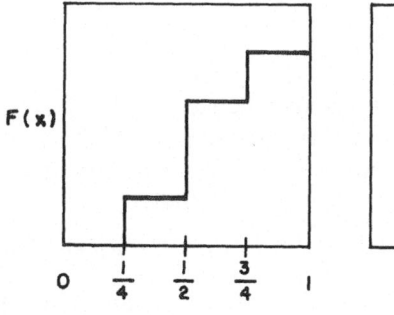

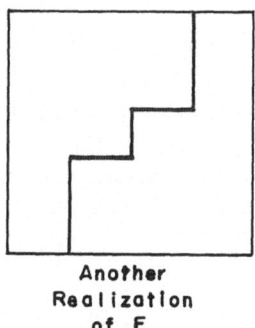

Another
Realization
of F

# 13.3. Bayes Theorem for a Dirichlet Process

**Theorem.** *Let $\alpha$ be a finite measure on $\mathcal{Y}$, and let $D_\alpha$ be a prior distribution on $\mathcal{P}$ the family of probabilities $P$ on $\mathcal{Y}$. Let $t$ be an observation from $T$ according to $P$. The posterior distribution of $P$ given $t$ is $D_{\alpha + \delta_t}$, where $\delta_t Y = Y(t)$.*

PROOF. The joint probability $Q$ on $\mathcal{P} \times \mathcal{Y}$ is defined by

$$Q[Z(P) \times Y] = Q^{\mathcal{P}}(Z(P)Q_{\mathcal{P}}^{\mathcal{Y}}Y) = D_\alpha(Z(P)PY).$$

The marginal distribution on $\mathscr{P}$ is $D_\alpha$, the marginal distribution on $\mathscr{Y}$ is $\alpha Y/\alpha 1$.

If $B_1, \ldots, B_k$ is a partition of $T$, then $P(YB_i)/PB_i$ is independent of $PB_1, \ldots, PB_k$ when $P$ is a Dirichlet process. (Since $P(YB_i)$ is the limit of a linear combination of disjoint sets $B_{ij}, \sum_j B_{ij} = B_i$ and $\{PB_{ij}/PB_i\}$ are Dirichlet $\{\alpha(B_{ij})/\alpha B_i\}$ independent of $B_i$.)

Thus if $Z(P)$ is of form $Z(PB_1, \ldots, PB_k) = Z_k$ say

$$D_\alpha[Z(PB_1, \ldots, PB_k)PY] = \sum_{i=1}^{k} D_\alpha[Z(PB_1, \ldots, PB_k)P(B_i)]D_\alpha[P(YB_i)/PB_i]$$

$$= \sum D_\alpha[Z(PB_1, \ldots, PB_k)PB_i]\alpha(YB_i)/\alpha B_i$$

If          $t \in B_i, D_{\alpha + \delta_t} Z_k = D_{\alpha B_1, \ldots, \alpha B_i + 1, \ldots, \alpha B_k} Z_k$

$$= D_\alpha(Z_k PB_i)/D_\alpha(PB_i)$$

Thus          $$D_{\alpha + \delta_t} Z_k = \sum B_i D_\alpha(Z_k PB_i)/D_\alpha PB_i$$

$$QD_{\alpha + \delta_t}(Z_k \times Y) = Q(Y\sum B_i D_\alpha(Z_k PB_i)/D_\alpha PB_i)$$

$$= \sum \alpha(YB_i)D_\alpha(Z_k PB_i)/\alpha B_i.$$

Thus $QD_{\alpha + \delta_t}(Z_k \times Y) = Q(Z_k \times Y)$ whenever $Z = Z_k$ depends only on $\{PB_i\}$.

Taking limits, the result holds for all $Z$, and so $D_{\alpha + \delta_t}$ is the posterior distribution of $P$ given $t$.          □

## 13.4. The Empirical Process

The limiting case of the Dirichlet $D_\alpha$ occurs when $\alpha \equiv 0$; this corresponds to the prior density $1/p(1 - p)$ for a binomial parameter $p$, which is not unitary. It is very difficult to imagine generating $P$ from $D_0$. The conditional distributions of $P$ and $t_{n+1}$ given $t_1, \ldots, t_n$ are nice and simple:

(i) $P_n Y = P[Y(T_{n+1})|t_1, \ldots, t_n] = (1/n)\sum Y(t_i)$, the empirical distribution over $\{t_i\}$.

(ii) $P|t_1, \ldots, t_n$ is Dirichlet $D_{\sum \delta_{t_i}}$; thus $[P(t_1), \ldots, P(t_n)]$ is Dirichlet $D_{1,1,\ldots,1}$ and $P$ is a discrete distribution carried by the observed sample points $t_1, \ldots, t_n$. See Hartigan (1971).

We are in the embarrassing position of declaring $t_{n+1}$ to be surely equal to one of the previous sample points; carried back, this would imply $t_n = t_1$ with probability 1, which is dull. We can pretend to the surprised at each new observation $t_{n+1}$ which is not equal to $t_1, \ldots, t_n$—after all events of probability zero do occur; but our credibility may be weakened by our always insisting that the next observation is just one of the previous ones, and our always being surprised!

To do error analysis of a parameter $Y(P)$ estimated by $Y(P_n)$, we compute the distribution of $YP$ where $P \sim D_{\sum \delta(t_i)}$. For example, if $Y(P_n) = (1/n)\sum t_i$,

$Y(P) = \int t dF$ is distributed as $\sum p_i t_i$ where $\mathbf{p} \sim D_1$; $P(\sum p_i t_i) = \bar{t}$, $\mathrm{var}(\sum p_i t_i) = \sum (t_i - \bar{t})^2 / n(n+1)$. This procedure gives approximate error behavior for any statistic based on the empirical distribution; it works best, when as here for the mean, the excessive discreteness is smoothed out by the statistic.

## 13.5. Subsample Methods

Consider the following competitors of the empirical process for generating posterior distributions of a functional $Y(P)$ estimated by $Y(P_n)$.

(i) Subsamples: Select a random subsample $t_{i_1}, \ldots, t_{i_r}$ of $t_1, \ldots, t_n$ where the $i^{th}$ observation lies in the subsample with probability $1/2$; regard $k$ random subsample values $Y(P_n)$ as a random sample from the posterior distribution of $Y(P)$. See Hartigan (1969).

(ii) Jackknife: Divide the sample into disjoint groups of size $k$ (randomly say). Define the $i^{th}$ pseudo-value by

$$Y_i = \frac{n}{k} Y(t_1, \ldots, t_n) - \left( \frac{n}{k} - 1 \right) Y(t_1, \ldots, t_n \text{ less } i^{th} \text{ group})$$

Act as if $Y(P)$ is a location parameter and $\{Y_i\}$ is a sample from $N(Y(P), \sigma^2)$. See Tukey (1958).

For example, let $n = 50$ and suppose $Y(P)$ denotes the correlation of a bivariate distribution. For the empirical process, sample $p_1, \ldots, p_{50}$ from $D_1$, and recompute the correlation on the data values weighted by $p_i$; obtain 3 such values. For subsamples, select 3 random subsamples each of size roughly 25. Do jackknifing with group size 25; if $r_1$ and $r_2$ are the correlations on the groups, the pseudo values are $2r - r_1, 2r - r_2$; the values $2r - r_1, 2r - r_2, 2r - (r_1 + r_2)/2$ are regarded as a random sample from the posterior distribution.

Each of the techniques gives 3 values, which divide the line into four intervals; in 100 repetitions, the true correlations lay in the four intervals as follows:

|                    | Bivariate normal $\rho = .95$ | | | | Mixture of normals | | | |
| ------------------ | -- | -- | -- | -- | -- | -- | -- | -- |
| Expected           | 25 | 25 | 25 | 25 | 25 | 25 | 25 | 25 |
| Empirical process  | 31 | 25 | 22 | 22 | 37 | 28 | 19 | 16 |
| Subsamples         | 31 | 23 | 28 | 20 | 33 | 23 | 27 | 17 |
| Jackknife          | 28 | 29 | 21 | 21 | 28 | 27 | 25 | 21 |

That's a bit nasty! By what accident could the humble *ad hoc* Jackknife beat such delightful Bayesian trickery? Hartigan (1975) shows that the asymptotic inclusion probabilities are correct for the various techniques.

## 13.6. The Tolerance Process

If $t_1, \ldots, t_n$ form a sample from a continuous distribution function $F$, and if $t_{(1)}, \ldots, t_{(n)}$ denote the order statistics, then $\{t_{(k-1)} < t_{n+1} < t_{(k)}\}$ is a tolerance interval for $t_{n+1}$ of size $1/(n+1)$; that is, $P[t_{(k-1)} < t_{n+1} < t_{(k)}] = 1/(n+1)$, averaging over all $t_1, \ldots, t_n, t_{n+1}$. After all, why should $t_{n+1}$ be any particular place in the ordered sample of $t_1, t_2, \ldots, t_n, t_{n+1}$?

The tolerance process defines $t_{n+1}$ given $t_1, \ldots, t_n$ to be such that

$$P[t_{(k-1)} < t_{n+1} < t_{(k)} | t_1, \ldots, t_n] = \frac{1}{n+1}.$$

More detailed probability statements are made as evidence accumulates. The joint distribution of $t_{n+1}, t_{n+2}$ given $t_1, \ldots, t_n$ may be computed by combining the $t_{n+1} | t_1, \ldots, t_n$ with $t_{n+2} | t_1, \ldots, t_n, t_{n+1}$; more generally $t_{n+1}, t_{n+2}, \ldots, | t_1, \ldots, t_n$ has a certain joint distribution. Obviously, $PA = \lim(1/n)\sum\{t_i \in A\}$, so the distribution of $P$ may be obtained from the distribution of $t_{n+1}, t_{n+2}, \ldots$; the distribution of $P$ is just that

$$F(t_{(1)}), F(t_{(2)}) - F(t_{(1)}), \ldots, F(t_n) - F(t_{(n-1)}), 1 - F(t_{(n)})$$

is $D_1$, or $F(t_1), \ldots, F(t_n)$ is a random sample from $U(0, 1)$.

$P[t_{(k)} < \text{median} < t_{(k+1)} | t_1, \ldots, t_n] = \binom{n}{k} 2^{-n}$ reproduces non-parametric confidence intervals for the median.

Here the probability space on which the distribution of $P$ is defined changes as evidence accumulates; Hill (1968) shows that no unitary probability on $P$ and $Y_1, \ldots, Y_n$ will reproduce these conditional probabilities, but Lane and Sudderth (1978) show that a finitely additive probability $P$ exists which produces these conditional probabilities.

## 13.7. Problems

E1. For observations 5, 7, 10, 11, 15 compute 50% confidence intervals for the median using the empirical process, subsamples, jackknife with group size 1, and the tolerance process.

P1. For $\mathbf{p} \sim D_1$, show that $\sum p_i X_i$ has skewness of opposite sign to that of $\mu - \bar{X}$. Thus if the $X$'s are positively skew, $\sum p_i X_i$ tends to be less than $\bar{X}$, but $\mu$ tends to be greater than $\bar{X}$.

P2. Let $t = \sum a_i X_i$ where $a_i = [(Z_i - (1/n)\sum Z_i)/Y] + (1/n)$, the $Z_i$ are independent $N(0, 1)$, and $Y$ is independent $[n\chi_{n-1}^2]^{1/2}$. If $\{X_i\}$ is a sample from $N(\mu, \sigma^2)$, and the prior density for $(\mu, \sigma^2)$ is $1/\sigma^2$, show that $t | X_1, \ldots, X_n$ and $\mu | X_1, \ldots, X_n$ have the same distribution.

P3. Let $X_1, \ldots, X_n$ be independent and symmetrically distributed about $\theta$. Let $Y_1, \ldots, Y_{2^n-1}$ denote the ordered means of the $2^n - 1$ subsets of $X_1, \ldots, X_n$. Show that $P_\theta(Y_k < \theta < Y_{k+1}) = 2^{-n}$, $1 \le k \le 2^n - 1$.

E2. If $X_{n+1}|X_1, \ldots, X_n$ is such that $P(X_{(k)} \leqq X_{n+1} \leqq X_{(k+1)}|X_1, \ldots, X_n) = 1/(n+1)$, find $X_{n+1}, X_{n+2}|X_1, \ldots, X_n$.

## 13.8 References

Blackwell, David and MacQueen, James B. (1973), Ferguson distributions via Polya urn schemes, *Annals of Statistics* **1**, 353–355.

Ferguson, T. S., (1973). A Bayesian analysis of some non-parametric problems, *Annals of Statistics* **1**, 209–230.

Hartigan, J. A. (1969), Use of subsample values as typical values, *J. Am. Stat. Ass.* **104**, 1303–1317.

——(1971), Error analysis by replaced samples, *J. Roy. Statist. Soc.* B **33**, 98–110.

——(1975), Necessary and sufficient conditions for asymptotic joint normality of a statistic and its subsample values. *Annals of Statistics*, **3**, 573–580.

Hill, Bruce M. (1968), Posterior distributions of percentiles: Bayes theorem for sampling from a population, *J. Am. Stat. Ass.* **63**, 677–691.

Lane, David A. and Sudderth, William D. (1978), Diffuse models for sampling and predictive inference, *Annals of Statistics* **6**, 1318–1336.

Neveu, J. (1965), *Mathematical Foundations of the Calculus of Probability*, San Franciso: Holden-Day.

Tukey, J. W. (1958), Bias and confidence in not-quite large samples, *Ann. Math. Statist.* **29**, 614.

# Author Index

# Subject Index

## A

Absolute distance   48
Admissibility   56−62, 63, 75, 76, 87,
    97, 98, 104, 105
  of Bayes decisions   x, 56−62
  various definitions of   x, 61−62
Analogy, Keynes uses   2
Approximable, mean-   35
  square-   35
Approximating sequence   16, 38
Asymptotic normality,
  crude demonstration   xii, 108
  examples   74−79
  martingale sequences   xii, 113−115
  of posterior distributions   xii, 107−118
  pointwise   xii, 111, 112
  regularity conditions   xii, 108, 109
Autoregressive process   102, 117
Axioms   ix, 14−22
  Kolmogorov's   5, 10, 15
  of conditional probability   23, 24

## B

Baire functions   40, 41
Baranchik's Theorem   xi, 84−86
Bayes   13
  decisions   57−62, 75

definition of probability   6
estimates   xi, 63, 86−87, 90, 92
postulate   2
robustness of methods   xii, 119−126
theorem   x, 30
theory   iii
unbiased tests   xi, 65−66, 75
Bayesian law of large numbers   36
Behrens-Fisher   81, 82
beta priors   76, 104
Bets   6, 9
Binomial,
  admissibility   61, 104
  asymptotics   116, 117, 126
  conditional probability for   x, 31−32
  convergence   x, 38−39
  exponential family   73
  methods   76−78
  priors   xi, 76−79

## C

Chisquares   93, 94
Clusters, multinomials with   101−102
Coherence   6
Collectives   3, 4
Complete Bayesian   101
Complete class   57
Complexity   4

# X

# Y